[GEEKED AT BIRTH.]

LEARN:

DIGITAL ANIMATION	**GAME PROGRAMMING**
DIGITAL ART AND DESIGN	NETWORK SECURITY
DIGITAL VIDEO	NETWORK ENGINEERING
GAME DESIGN	SOFTWARE ENGINEERING
ARTIFICIAL LIFE PROGRAMMING	WEB ARCHITECHTURE
COMPUTER FORENSICS	ROBOTICS & EMBEDDED SYSTEMS

University of Advancing Technology

UAT

Learn. Experience. Innovate.

IM Geek PH

).10 PWR: 110

You can talk the talk. Can you walk the walk? Here's a chance to prove it. Please geek responsibly. www.uat.edu > 877.UAT.GEEK

Make:
technology on your time®

Volume 16

ON THE COVER: Watch your back! Two of the most infamous spies in history, Black and White, scheme in true maker fashion. Look for them sneaking throughout this issue. Illustration by Sam Viviano; SPY vs. SPY ™ & © E.C. Publications, Inc.

Columns

➕ **makezine.com**
Visit for story updates and extras, Weekend Project videos, podcasts, forums, the Maker Shed, and the award-winning MAKE blog!

Vol. 16, Nov. 2008. MAKE (ISSN 1556-2336) is published quarterly by O'Reilly Media, Inc.in the months of March, May, August, and November. O'Reilly Media is located at 1005 Gravenstein Hwy. North, Sebastopol, CA 95472, (707) 827-7000. SUBSCRIP-TIONS: Send all subscription requests to MAKE, P.O. Box 17046, North Hollywood, CA 91615-9588 or subscribe online at makezine.com/offer or via phone at (866) 289-8847 (U.S. and Canada); all other countries call (818) 487-2037. Subscriptions are available for $34.95 for 1 year (4 quarterly issues) in the United States; in Canada: $39.95 USD; all other countries: $49.95 USD. Periodicals Postage Paid at Sebastopol, CA, and at additional mailing offices. POSTMASTER: Send address changes to MAKE, P.O. Box 17046, North Hollywood, CA 91615-9588. Canada Post Publications Mail Agreement Number 41129568. CANADA POSTMASTER: Send address changes to: O'Reilly Media, PO Box 456, Niagara Falls, ON L2E 6V2

PLAY UNFAIR:
Tap into a tricky drawer that opens only when the right moves are made.

Make: Projects

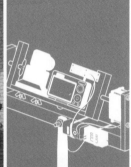
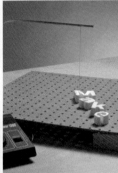
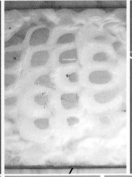

YOU SHOULD HAVE A HOUSE WITH AN A/C UNIT ATTACHED TO IT, NOT THE OTHER WAY AROUND.

With air conditioners, smaller is better. And Johnson Controls was the first to offer YORK® home A/C units that use ingenious micro-channel technology to produce more cool in less space, while still efficiently saving energy. Just another way we're making this world a more comfortable, safe, and sustainable place. Want to make your world better? We'll help you figure it out at ingenuitywelcome.com

Johnson Controls

INGENUITY WELCOME

Make:

Volume 16

technology on your time®

I, ROBOT:
Marque Cornblatt decided open source was the only way to go with his wireless, rolling, remote control robot, Sparky 2.

Make:
technology on your time

EDITOR AND PUBLISHER
Dale Dougherty
dale@oreilly.com

EDITOR-IN-CHIEF
Mark Frauenfelder
markf@oreilly.com

CREATIVE DIRECTOR
Daniel Carter
dcarter@oreilly.com

MANAGING EDITOR
Shawn Connally
shawn@oreilly.com

DESIGNERS
Katie Wilson
Alison Kendall

ASSOCIATE MANAGING EDITOR
Goli Mohammadi

PRODUCTION DESIGNER
Gerry Arrington

SENIOR EDITOR
Phillip Torrone
pt@makezine.com

PHOTO EDITOR
Sam Murphy
smurphy@oreilly.com

PROJECTS EDITOR
Paul Spinrad
pspinrad@makezine.com

ONLINE MANAGER
Tatia Wieland-Garcia

COPY CHIEF
Keith Hammond

ASSOCIATE PUBLISHER
Dan Woods
dan@oreilly.com

STAFF EDITOR
Arwen O'Reilly Griffith

CIRCULATION DIRECTOR
Heather Harmon

EDITORIAL ASSISTANT
Laura Cochrane

ACCOUNT MANAGER
Katie Dougherty

EDITOR AT LARGE
David Pescovitz

MARKETING & EVENTS MANAGER
Rob Bullington

MAKE TECHNICAL ADVISORY BOARD
Kipp Bradford, Evil Mad Scientist Laboratories, Limor Fried, Joe Grand, Saul Griffith, William Gurstelle, Bunnie Huang, Tom Igoe, Mister Jalopy, Steve Lodefink, Erica Sadun

PUBLISHED BY O'REILLY MEDIA, INC.
Tim O'Reilly, CEO
Laura Baldwin, COO

Visit us online at makezine.com
Comments may be sent to editor@makezine.com

For advertising inquiries, contact:
Katie Dougherty, 707-827-7272, katie@oreilly.com

For event inquiries, contact:
Sherry Huss, 707-827-7074, sherry@oreilly.com

Customer Service cs@readerservices.makezine.com
Manage your account online, including change of address at:
makezine.com/account
866-289-8847 toll-free in U.S. and Canada
818-487-2037, 5 a.m.–5 p.m., PST

Contributing Editors: Gareth Branwyn, William Gurstelle, Mister Jalopy, Brian Jepson, Charles Platt

Contributing Artists: Matt Blum, Jordan Bunker, Michael T. Carter, Amy Crilly, Nick Dragotta, Dustin Fenstermacher, Julian Honoré, John Keatley, Bob Knetzger, Jonathan Koshi, Garry McLeod, Pat Molner, Branca Nitzsche, Bill Oetinger, Cody Pickens, Nik Schulz, Damien Scogin, Jen Siska, Robyn Twomey, Sam Viviano

Contributing Writers: Dan Albert, Thomas J. Arey, John Baichtal, Dan Bassak, Ryan Beacom, Jake Bronstein, Bill Byrne, Kenny Cheung, Marque Cornblatt, Dick DeBartolo, Ken Delahoussaye, Brian Dereu, Cory Doctorow, Scott Driscoll, Simon Quellen Field, Daniel Gentleman, Mike Golembewski, Brian Graham, James Grant, Ross Griffith, Saul Griffith, Donald A. Haas, Mitchell Heinrich, Jess Hemerly, Wayne Holder, Lisa Katayama, Bob Knetzger, Richard Langevin, Andrew Lewis, Greg MacLaurin, Brian McNamara, Erico Narita, Ken Olsen, John Edgar Park, Michael H. Pryor, Paulo Rebordão, Eric Rosenthal, Donald E. Simanek, David Simpson, Star Simpson, R.U. Sirius, Eric Smillie, Peter Smith, Bruce Stewart, Donna Tauscher, Jason Torchinsky, Bryant Underwood, Megan Mansell Williams, Edwin Wise, Frank Yost, Lee D. Zlotoff

Bloggers: Jonah Brucker-Cohen, Collin Cunningham, Mike Dixon, Luke Iseman, Kip Kedersha, Patti Schiendelman, Becky Stern, Marc de Vinck

Interns: Kris Magri (engr.), Harry Miller (engr.), Lindsey North (projects), Meara O'Reilly (projects), Ed Troxell (photo)

solder by numbers™

Do you remember paint by numbers?
If you could paint, you could make a great picture.

Solder By Numbers™ is the same concept but for electronics.

With no previous experience you will simultaneously build electronic circuits and a circuit of new worldwide friends.

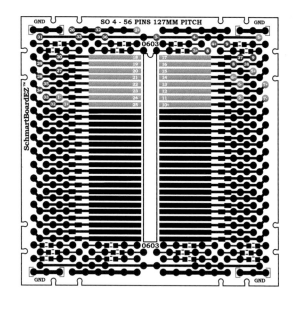

Solderbynumbers.com is the first social networking site for electronics enthusiasts. It will greatly enhance the experience of everyone from the electronics novice to the seasoned pro.

Sign up for your free account today!
www.SolderByNumbers.com

Contributors

Sam Viviano (Cover, special section opener, and various other spy illustrations) was born sometime in the last century to an anonymous couple in Detroit. After graduating *summa cum laude* from the University of Michigan, he moved to New York, where he quickly became impoverished. He was saved from a life of crime by the editors of *MAD*, who assigned him his first cover in 1980. In no time at all, he was one of the Usual Gang of Idiots, producing movie and television satires, phony ads, and, of course, more covers. Finding no other way to stem his seemingly endless output of bad illustration, the editors asked him to become the art director of *MAD* in 1999, a position he still sleeps through today. Viviano lives in New York with his wife and daughter, who (like his parents) would prefer to remain anonymous.

Brian Dereu (*Dead Drop Device*) has been self-employed now for more than 12 years and has never looked back. "It took me a while to develop the discipline to work for myself," he says, "instead of just going fishing." He runs a small manufacturing business in rural Missouri where his day consists mainly of lathe and milling machine work (when he's not fishing). Even after hours, he can be found tinkering in the shop or on the computer working on his newest venture (spy-coins.com). Brian is a family man, with a wife and 9-year-old twin boys at home. His wife, Sonya, home-schools the twins, and the entire family is active at their Christian church.

Known to his wife, Linda, as Mr. Experimentation, **David Simpson** (*G-Meter and Altimeter, Covert Wireless Listening*) was inspired as a child by his Uncle Marv, "a hot rodder from the 50s. He would heat up plastic car models in the oven and squash them into fascinating, realistic wrecks." Since then, David has built "a lot of off-beat things from found objects," but his real passion is aviation and aerospace education, which "has been under my skin for a while." He finally got his private pilot's license a few years ago, and is "always working on developing new, immersive, hands-on educational experiences for cadets of the Civil Air Patrol" in New Jersey.

Star Simpson (*Star Bust*) started her career at age 3, when her mother told her not to mix electricity and water. "So I stuck my fist under running water," she recounts, "and grabbed the contacts of a night light plugged into the wall next to the sink with my other hand. I got a 120V jolt and found out my mom wasn't making things up. I've thought electricity was awesome ever since." Currently a world traveler with a penchant for photography, surfing, and tinkering, Simpson grew up in Hawaii and attended MIT; she's now working on an outrigger canoe to sail between islands in Hawaii. Her favorite tool is a homemade welder made from salvaged microwave oven transformers, but she also thinks "a nice TIG is a good thing."

When **Andrew Lewis** (*Secret Chessboard Drawer, Self-Destructing Object, USBattery*) isn't "hacking away on a problem connected to computational complexity," he's most likely "working in my lab on a 3D scanner or some other hardware-related project." The computer engineer, author, and Ph.D. student lives in a small village surrounded by farmland, not far from the urban center of Staffordshire, England. Andrew is a steampunk fan with a decidedly madcap attitude to making: "I think part of the fun is when something goes wrong and provides you with a new challenge ... unless you're on a tight schedule!"

Dan Bassak (*Roller Skate Toe Stop*) hails from northeastern Pennsylvania. Growing up on a dairy farm, Dan learned early on that he could freak out his entire neighborhood with his inventions. His parents grew accustomed to the sound of explosions from his bedroom and the sight of the local police on the front porch. Lucky for the planet, Dan channeled his talents and energy toward the greater good as he got older. Dan is also known for his excellent culinary skills, most notably his bread- and sausage-making abilities. Presently employed as a biomedical engineer, Dan resides with his tolerant wife, Trina, and their dog, Abby.

Photograph of Star Simpson by Jeff Lieberman

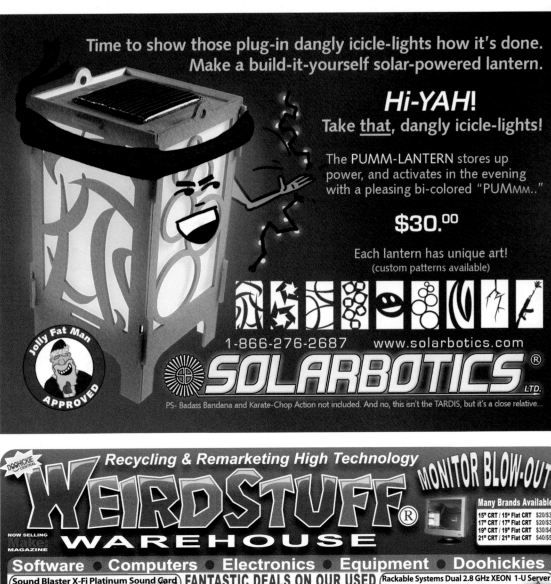

The Visible Hand

As I write this, there is panic on Wall Street despite Washington's $700 billion rescue attempt. The crisis is not contained by U.S. borders, but extends to Europe and Asia. Like many people, I'm incredulous. How could this happen?

Wall Street hired the best and the brightest, paid them handsomely, and gave them unlimited resources and technology. It turns out they were building enormously complicated castles made of sand. A great wave washed them away, astounding all the smart people who devoted their lives to speculation, not production. Their models based on historical data predicted future profits, not collapse. Few people saw this coming until it hit.

"It was the triumph of data over common sense," said reporter Adam Davidson on the excellent episode of *This American Life* called "The Giant Pool of Money." Economist Michael Lehmann in the *San Francisco Chronicle* called it "the triumph of ideology over common sense." It's obvious both common sense and the common man have taken a beating.

It's hard to stomach that our government must bail out Wall Street. It really means we've bet our future on the same people who created the present situation. To paraphrase a joke I've heard: It's like going to a casino in Vegas and rooting for the house. One *New York Times* reader expressed the frustration that many feel: "Why can't we take half of the $700 billion and just *build* something?"

These events shake our belief that free markets work to the benefit of all. The fundamental tenet of capitalism is the "invisible hand": Adam Smith wrote that "by pursuing his own interest [each person] frequently promotes that of the society." This year, Nobel Prize-winning economist Joseph Stiglitz said: "In this sense, the fall of Wall Street is for market fundamentalism what the fall of the Berlin Wall was for communism — it tells the world that this way of economic organization turns out not to be sustainable."

A headline in the *Christian Science Monitor* says: "With finance crisis, hands-off era over." Government will need to be more assertive in regulating Wall Street. But I think it goes beyond that. I wonder if we, as individuals, have been living in our own era of hands-off. Have Americans become so disengaged that we've become dependent on some invisible force to provide what we need? Have we gotten used to leaving important matters to experts, until they turn out to be wrong?

Isn't it time for us to become hands-on again?

We, the people, face enormous challenges. Apart from the economic mess, we know fundamental changes are coming because of global warming. Our dependence on fossil fuels is not sustainable. Change is coming, whether we want it or not.

Better we meet the challenges head-on rather than hide. *New York Times* columnist Thomas Friedman

The DIY mindset must again become an essential life skill.

summed it up: "We need to get back to making stuff, based on real engineering not just financial engineering. We need to get back to a world where people are able to realize the American Dream — a house with a yard — because they have built something with their hands, not because they got a 'liar loan.' ... The American Dream is an aspiration, not an entitlement."

We have to believe it starts with each of us — not some faceless government or corporate bureaucracy. It's time for us, individually and working together in business, to reconsider what it means to be productive, not just profitable. It's time for us to reengage in how our government sets priorities for education, health care, housing, and transportation.

The DIY mindset celebrated in this magazine must again become an essential life skill, rooted once again in necessity and practicality. Our future security lies in knowing what we're capable of creating, and how we can adapt to change by being resourceful.

A challenge this great can bring out the best in us. We need everyone, because every person has something to contribute. We need a showing of all hands.

Dale Dougherty is the editor and publisher of MAKE and CRAFT magazines.

BY CORY DOCTOROW

Selectable Output Control

Chances are, you haven't heard of "selectable output control" (SOC), a proposed digital TV technology that would allow broadcasters to tag their content with a list of devices that are allowed to play it. That's because it's an insane idea.

Picture this: you power up your home theater, a complex network of game consoles, A/V switchers, cable boxes, PVRs, DVD players, 5.1 speakers, amps, a home theater PC, and a monitor or projector. After locating the correct remote, you start surfing through channels. All good. But when you hit MTV, the gorgeous, perfectly balanced sound stops.

Why? Because MTV doesn't want you digitizing the songs that accompany its music videos, so it sends a digital "flag" that disallows high-end audio on equipment that doesn't contains digital rights management (DRM). Your beautiful hand-built tube amp certainly isn't compatible, so if you want sound while watching MTV, you've got to turn on the tiny internal speakers that came with your TV.

You surf on up the dial (get up again and turn off the internal speakers), and flip to HBO. Your screen goes dark. That's because HBO is showing a movie that has been flagged as "no analog" — which means your beautiful 42" plasma display won't work because you connected it via the composite analog video cables coming off the back of your A/V switcher, rather than via the DRM-locked HDMI output.

To watch the movie, you'll need to move the entire shelving unit (remember to take down the family photos first, doofus, otherwise you risk shattering the glass if they tip over), disconnect the analog cables, dig around in the garage to find the HDMI cable that came with the TV (or was it the cable box?), and rewire your set.

One more channel up the dial and the screen goes dark *again*. Google around for a while, and you discover that some kid in the Ukraine published a class break last month for HDCP, the anti-copying crap in HDMI, that makes it possible to record HDCP content using "unauthorized" technology. So here on Showtime, they've restricted HDCP as well as analog. You're going to need DVI for this one. You grab a little Maglite and peer hopefully at the inputs on the plasma. No DVI. Now what? Buy a new TV?

The dead hand of copyright is in the guts of your inventions.

This is where SOC crosses the line from totally objectionable to totally insane. With SOC, it doesn't matter if you're careful to buy only "approved" technologies and set them up in the "approved" manner. Because no matter how you set your stuff up *today*, the signal can be altered to prohibit access *tomorrow*, if someone, somewhere, figures out how to do something naughty with a device you have the misfortune of owning.

You haven't heard of SOC because, in 2003, the FCC told broadcasters and cable operators that they weren't allowed to use it. But, like a bad Hollywood sequel, it's back. The Motion Picture Association of America has petitioned the FCC for the right to turn on SOC for new-release movies. It promises it won't use SOC for other purposes, but you can bet its use will expand and expand.

SOC isn't just a cable proposal, it's an entire philosophy that's the antithesis of making. It's a philosophy that says that the dead hand of the original manufacturer will be an immortal presence in every device you own, yanking out the wires that it objects to, turning the dials to suit its needs. If you've ever done something with a device that the manufacturer didn't intend (or wouldn't like), you can appreciate how bad an idea this is.

Copyright has its place. I think that it's totally legit to propose that someone who makes a creative work should be allowed to control the circumstances under which companies can sell copies of it. But since when does copyright give a creator the right to tell you which wires to plug into your TV?

» The Electronic Frontier Foundation — which fought SOC in 2003 — is fighting it again, and you can help out at eff.org/issues/digital-video.

Cory Doctorow lives in London, writes science fiction novels, co-edits Boing Boing, and fights for digital freedom.

Tales of hardware stores, wind generators, and pine derby cars.

✉ My son and I were in our local independent hardware store, Jackson's Hardware in San Rafael, Calif., shopping for parts for the compressed air rocket (which is fantastic!) [*Volume 15, page 102, "Compressed Air Rocket"*]. There was another father/son combo in the store at the same time buying parts for the rocket. I happened to speak to the store manager and told him that there was another party buying the exact same stuff for their rocket project. He had not heard of MAKE but said, "We should carry that magazine." I told him that if a project was popular he could even put together a kit of parts to make it easy on us parents and tinkering goobers.

So, my idea is to get your mag into more independent hardware stores, and it will do us all good. Sure, if Home Depot wants it, give it to them, but make sure the indy shops get some love.

Anyhow, just an idea. Thanks for the great mag. I look forward to it more than any other. My son loves the compressed air rocket. It is really fun.

—*Alex Giedt*
San Rafael, Calif.

✉ I have been an avid reader of MAKE since the very beginning, back in 2005, and I would like to thank you for keeping me inspired to continue with my creative ventures. All of my science projects in recent memory have been based on projects you've featured in MAKE. For example, last year I compared the efficiency of various wind turbine designs, inspired by the article "Wind Powered Generator" in Volume 05 [*page 90*].

MAKE is a very useful reference for any kind of project I'm working on. I often find myself reading the online PDF more often than the physical copy; it's very convenient to have it right there on my computer. Your blog is also great; it keeps me occupied while waiting for the next issue, and it's updated often enough that I can check on it several times a day without getting bored, unlike most blogs which, at most, update once every day or so. I'm looking forward to future issues, keep up the great work!

—*Jacob Simmons*
Lake City, Fla.

✉ Regarding "Interstellar Visions," Volume 15, page 21: your reporter is too credulous. While concentrated moonlight may have medical value for werewolves, its only effect on humans is to lighten their wallets. The Interstellar Light Applications collector may be technologically impressive and artistically interesting, but scientifically and medically speaking it is indeed in the middle of a desert. You do your readers a disservice to suggest otherwise.

—*Eric Johnson*
Minneapolis, Minn.

✉ I loved the "Model Wind Tunnel" article [*Volume 15, page 143*]. My dad and I made a wind tunnel for the 5th grade science fair. The nice touch that we (probably he) figured out was to use dry ice and a beer can with holes punched in it in front of the air inlet to make little lines of dry ice fog that showed how different shapes were more turbulent than others. Mine was narrow and only worked for airfoils, not Pinewood Derby cars — though since my dad had the best set of tools, our scout troop's Pinewood Derby cars were all started in our basement.

—*Ben Smith*
San Francisco, Calif.

MAKE AMENDS

Frank Ford, author of several tips in MAKE, Volume 15, was referred to with the incorrect name. Additionally, Nate Ball wrote the *Design Squad* go-kart article on page 48 of the same issue, and we spelled his name wrong. We apologize for the mix-ups.

Photograph by David Olsen

DIY KITS + TOOLS + BOOKS + FUN

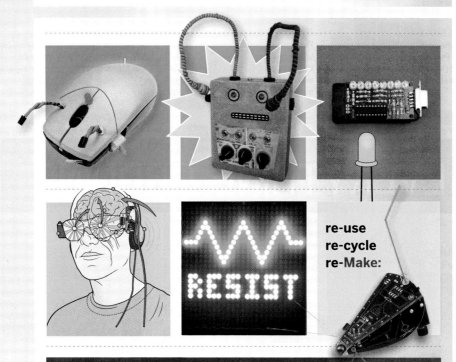

re-use
re-cycle
re-Make:

Welcome to the Maker Shed. In the following special holiday section, you'll discover just a few of the unusual and hard-to-find DIY projects available for purchase in our makershed.com store. Carefully screened by the editors and staff of MAKE magazine, Maker Shed kits, tools, books, and games are designed and produced by our favorite makers and small suppliers from around the world. Our goal is to offer life-enriching challenges and exploration through carefully curated science, tech, and crafting projects for a range of interests and experience levels. We hope you enjoy them!

HAPPY HOLIDAYS + PERMISSION TO PLAY!

The Value of a Good Hands-On Project

Editor and Publisher Dale Dougherty came by my desk the other day pointing at a *Newsweek* folded back to a chart that ranked retail winners and losers for the past quarter.

"Look what came in right behind gasoline stations on the high-growth list," he said, pointing to the circled chart. "Hobby, toy, and game stores." No one who owns a car will be surprised to see that gas stations top the growth chart, but hobby and game stores? We're in the midst of some gray economic times, and folks generally think of hobbies and games as discretionary pursuits, no?

Dale and I were intrigued by the chart because it mapped so closely to our own Maker Shed experience — a pronounced upswing in interest in kits. To be sure, part of our success is the result of a team of smart editors and staffers who've uncovered beautiful kits and projects that really resonate with our audience of inquisitive makers and science enthusiasts. However, I think the underlying data is telling us something important about ourselves and the kind of value we derive from a good hands-on project.

Perhaps it's the constructive distraction of focusing ourselves on something other than the recession, something where we have a reasonable chance of controlling the outcome. Maybe it's the satisfaction of picking up a new skill, dusting off an old one, or simply learning how something works (or doesn't). Maybe it's the memories that live long after the project is done.

And there's definitely something intrinsically satisfying about passing along skills — even the simplest of skills — to a younger maker. What kid doesn't enjoy a workbench, a few tools, and a good project on a rainy day?

Even though many of us are nixing the vacation we'd thought about, driving that funky clunker of a car for another year, or putting the bathroom remodel on hold, the basements, garages, and backyards of this planet are coming alive with experiments, tinkering, and the making spirit.

So this holiday season, whether you provision a project from recycled materials and repurposed

> The chart mapped so closely to our own experience: a pronounced upswing in interest in kits.

hardware lying around the house, or decide to buy a project kit from the Maker Shed (makershed.com) or somewhere else, give yourself and someone you care about the gift of making something together.

And if you're in a position and the spirit moves you, consider giving the gift of a science kit to a deserving school or teacher. They need your help more than ever before.

Dan Woods is associate publisher of MAKE and CRAFT magazines.

MADEONEARTH

Report from the world of backyard technology

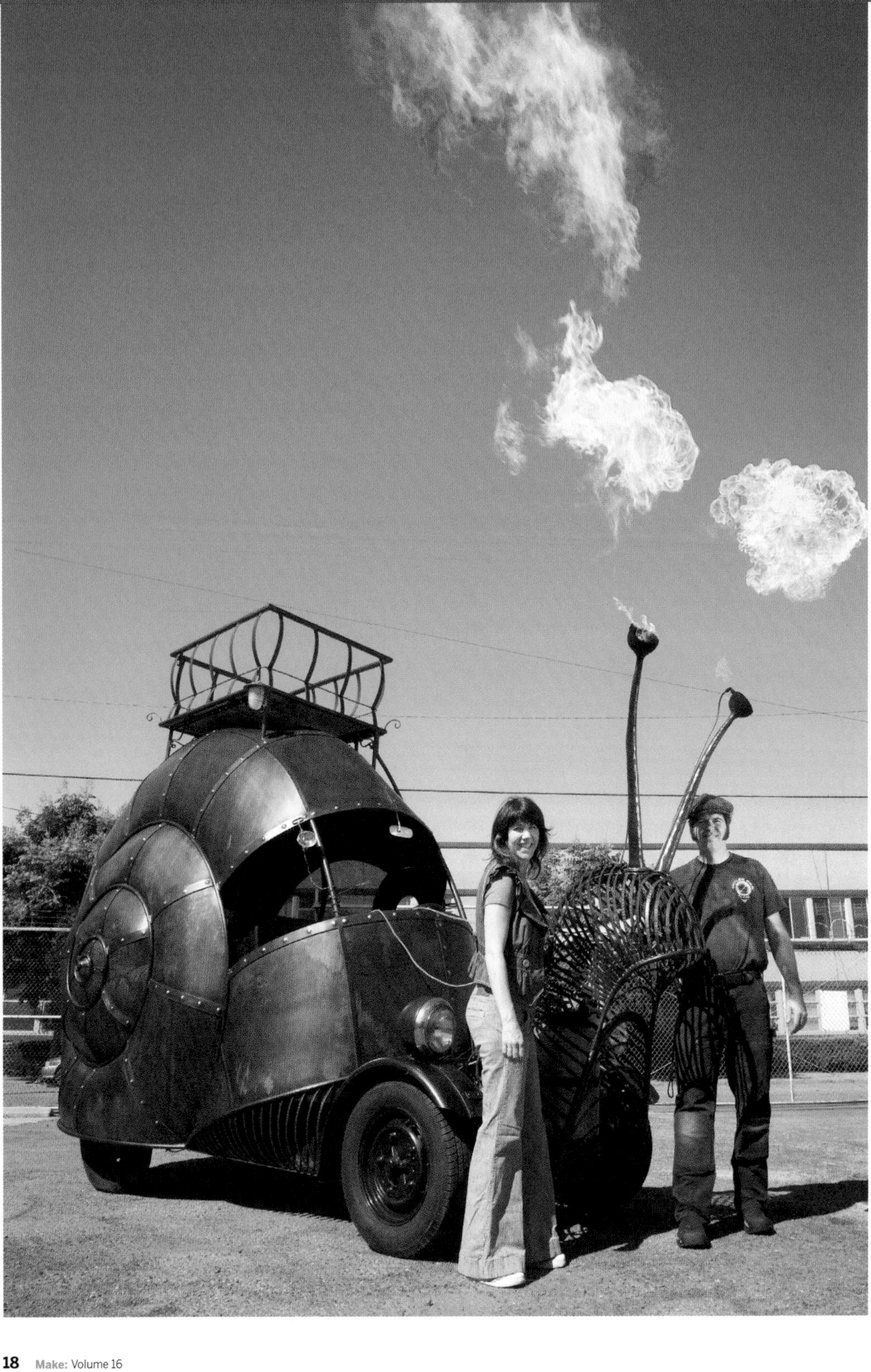

Photography by Branca Nitzsche

Out of Their Shell

Kyrsten Mate drives like a snail. No surprise, though. After all, her car *is* a snail. Dubbed the *Golden Mean*, it debuted at this year's Burning Man festival, where revelers eagerly rode in its glowing shell, or hugged the head with its flamethrower eyes. Every afternoon, Mate's 2-year-old daughter, Zolie, commandeered the cushioned interior for story time.

"Several years ago, I dreamt that I saw a car in the desert but the car was a giant snail," says Mate, a sound designer for films like *The Incredibles*.

And those are the kinds of dreams that stuff is made of. Fortunately, Mate and her husband, blacksmith **Jon Sarriugarte**, are well versed in weird rides. Their last collaboration was the *SS Alpha Fox*, a civil service vehicle converted into a fire-spewing spaceship straight out of 1960s science fiction.

The *Golden Mean* began life as a 60s-era Volkswagen Beetle — inexpensive and easy to transform with their "oil punk" aesthetic. Early plans for the shell called for fiberglass construction, but Sarriugarte wasn't eager to deal with the mess.

Fortunately, the scrap pile at Sarriugarte's hand-forged furniture company, Form & Reform, was heaping with possibility. They made an exoskeleton from steel, skinning it in sheet metal and perforated steel to allow a view of the outside world.

Midway through the build, the couple learned about the golden ratio of mathematics, sometimes called the golden mean. The number is found throughout nature, including the proportions of a snail's shell. They soon took a tape measure to their creation.

"We realized that we had unconsciously already been following the proportion a lot," Sarriugarte says. "That revelation has given me a whole new insight into how I'd like to make stuff inspired by nature."

The biomechanical buggy does seem alive. It boasts an air ride suspension that can pump the shell up and down. Other lively mods are in the works.

"We can't forget the snail trail," Mate reminds him. That would be a trickle of graywater collected from campsites when the *Golden Mean* hits the open road.   —*David Pescovitz*

>> ***Golden Mean* and More:** formandreform.com

Shake-o-Rama

Kiwi conceptual artist **D.V. Rogers** likes to shake it. To prove it, he dug a trench 6 feet down near the San Andreas Fault in Parkfield, Calif., population 18.

Into the void he lowered his very own hydraulic shake table, rescued from a defunct mining museum and set to jostle, its earthquake data captured by the U.S. Geological Survey, where Rogers is an artist in residence. Atop the hulking, three-ton slab, he attached an array of 10-foot, ⅝-inch steel rods that oscillate to the Earth's seismic hum.

The project, *Parkfield Interventional EQ Fieldwork* — the EQ stands for earthquake — started rumbling in the late 1990s, when the artist, based in Sydney, Australia, acquired the table. A couple of years and 3,000 hours of labor later, Rogers had a modular machine that moved when global earthquakes rattled.

An early installation caught the attention of USGS scientist Andy Michael, who suggested bringing the work to Parkfield, the most heavily monitored part of the globe, seismically speaking, known for withstanding a magnitude 6.0 or greater quake every 22 years or so (the last one was in 2004).

"Earthquakes are in my blood, I guess," says the cowboy hat-clad Rogers, who's from Napier, New Zealand, a port city on the North Island leveled by a magnitude 7.9 quake in 1931.

Using software developed by Stock Plum in the Netherlands and Dr. Geo Homsy in Alameda, Calif., *PIEQF* runs a Python script to isolate California data from the USGS monitoring network. Its I/O card drives a bank of relays that switch solenoid valves and open the gates to hydraulic actuators, allowing the entire table to move horizontally and vertically. The swaying rods above accentuate the table's movement and that of the ground below.

"The idea is to create a physical representation of a dynamic landscape," Rogers says at his characteristic lightning clip. "I'm *intervening* and introducing a human time scale, literally creating a reflector of the geologic time frame we reside in."

—*Megan Mansell Williams*

Earthquake Machine: allshookup.org/parkfield

Photograph by Scott Haefner USGS

Photograph by Gail Simpson

Avian Spacecraft

It wasn't your typical crash landing, but, then again, it wasn't your typical UFO, either. On a wet May day on the outskirts of Philadelphia, Gail Simpson and Aristotle Georgiades forklifted a 500-pound wooden flying saucer 20 feet up into a grove outside the Abington Art Center.

As Georgiades and a helper clambered on top to cable it to a tree, the rain came pelting down. "There was a windstorm, a rainstorm, and the temperature had dropped 20 degrees," says Simpson. "It was kind of nightmarish."

The UFO, on the other hand, is anything but. Made from a warm and familiar material — weathered pine boards left over from a turn-of-the-century barn — the sculpture isn't likely to inspire fear. And eight built-in birdhouses add an inviting touch.

The duo, who go by the name Actual Size Artworks, aimed to give the spacecraft "a really handmade, crazy-carpenter look," says Simpson, adding that this isn't a big departure from the Martian ships of the 1950s: "When you look at them in old posters and movies, they do look a little rickety."

Building the giant donut (the center of the alien ship) was equal parts high tech and handicraft for the two University of Wisconsin-Madison professors, who combined a CAD software design with a lot of fiddly sawing. "The program gives you an idea of size and angles but there's always human operator error, and the materials were all eccentrically sized," explains Simpson.

They built an armature of ribbed sections and skinned it with the boards. At 16 feet in diameter, it easily allowed Simpson to slip inside and spend a lot of time there threading bolts through.

"The scale is true to little green men," says Georgiades.

"It would comfortably house two small aliens," chimes in Simpson.

"Oh," counters Georgiades, "it would house four or five." Not to mention a whole flock of birds.

—*Eric Smillie*

≫ **Larger-Than-Life Sculpture:** actualsizeartworks.com

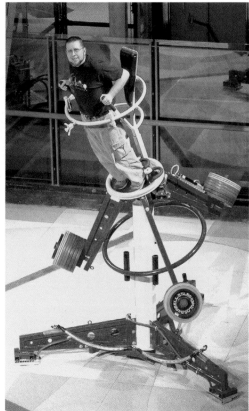

Gliding on Air

"Nostalgia on steroids" is how **Tom Luczycki** describes the process he used to build the Air Car. Recalling the personal hovercraft ads he saw as a child in magazines like *Boy's Life*, Luczycki set out to build an exhibit to delight children and remind adults of the one device they always wanted to build.

The original model used industrial air casters and a lightweight camp chair to float the rider. The second iteration replaced the camp chair with some additional framework and an aluminum race car seat that added several pounds of mass, but tons more cool.

Then came directional jets. The first version was just a single jet; the rider held a length of hose and operated a ball-valve, which allowed them to get moving pretty quick, but — here's the kicker — steering and orientation were impossible.

Luczycki's team wanted the air car to simulate the difficulties of maneuvering in space — weightlessness in 2D — so a more complex system was needed. They came up with an arrangement where the rider manipulates the car via a joystick and a lever switch.

Luczycki first got the maker bug when he worked on the maintenance crew at a paper mill in upstate New York. "When trying to meet a production quota at the end of the month, you saw the most insane, Goldbergian cobb-jobs to keep a machine running for just a few more days," he recalls. Once the deadline passed, he'd get to see the job done right. He learned that both methods are important, and need to be used together in the right proportions.

After studying engineering, Luczycki switched to fine arts and worked in an art foundry, first as a metal finisher, then as a large-scale sand molder.

Eventually, he landed his dream job as a designer and fabricator at the Detroit Science Center, where he built projects like the Air Car and the Tilt-O-Rama, a ride that elicits two kinds of reactions: the shock and horror of the passengers and the delight of their friends at seeing them in shock and horror.

"The best is when we would put in a teacher who was visiting the center on a field trip," he says. "The students went nuts." —*Bruce Stewart*

» **Tom's Stuff:** land-o-lizardo.blogspot.com

Photography by Tom Luczycki

Technozoology

"I'm in a bird phase," says **Ann Smith** of Providence, R.I. The birds in question evoke playfulness and scientific savvy with a meticulous arrangement of mechanical and electronic castoffs, including keyboard mylar and tiny gears from watches.

These fanciful avian miniatures are just the latest creations in her growing mechanical menagerie. Past enthusiasms include bugs, goats, cats, antelopes, frogs, squid, jellyfish, dinos, and more.

Smith's first robot-like figurine was a horse she made for an illustration assignment at the Rhode Island School of Design. She explains: "The topic of the project was technology and my idea was to make some sort of a Trojan horse of technology: technology being something people invite into their homes that eventually spreads and takes over."

Both natural and mechanical forms continue to inspire Smith's work, and the sources of her material are as diverse as the creatures themselves: sewing machines, clocks, computers, keyboards, phones, cameras, printers, typewriters, and other gadgets.

She starts her sculptures by focusing on the structural form of an animal, and then builds onto it, adding texture and detail. "A piece only feels complete when it's acquired a sense of life and personality," she says.

The distinctive personalities of her robotic figures have been used as illustrations for advertising and featured in prestigious magazines such as *Wired* and *Architectural Digest* (Germany). Her work is also in demand in shops and galleries nationwide.

But her new passion is breathing life into her creations through stop-motion animation. "They have always been alive and running around in my mind, and it is wonderful to see them actually come to life," she exudes. "Eventually I'd like to start adding narration and sound to the pieces."

In the meantime, her miniature zoo expands with more colorful and delicate birds. And, her imagination ever alert to possibility, a 3½-foot-long whale.

—*Donna Tauscher*

≫ **Ann Smith's Sculptures:** burrowburrow.com

Photography by Ann Smith

Spying for Kicks

In a panopticon, every prison cell is visible from a central guard tower. No one can get away with anything and so, in theory, everyone behaves. *Kalopticon*, on the other hand, a life-sized robot created by San Francisco artist **Kal Spelletich**, invites disorder, and toys with the notion of surveillance.

Spelletich was born in a hospital elevator. At 9, he began tinkering with a chemistry set. He took a construction job as soon as he could hold a hammer, and started rebuilding engines before he had a license. He's also an actor — think "TV guy" in the 1991 film *Slacker*. Oh yeah, and there's his whole art thing: degrees from the University of Iowa and the University of Texas, and lecturer at San Francisco State University.

The artist, whose interactive works include a burning bed that straps a participant in for a backward free-fall, unveiled *Kalopticon* at the *Close Calls: 2008* show in Sausalito, Calif. Spelletich used his own measurements, cast his own face, and dressed his creation in his own duds. But *Kalopticon* is rude. It kicks, jerks, and turns away just when

you lean in for a closer look. Machine malice isn't at work, though. Cameras in the robot's mouth and on the ceiling above feed real-time footage to TV monitors 30 feet away, where a joystick lets participants control *Kalopticon*'s flails.

The installation tickles the Big Brother nerve intentionally, Spelletich says, noting that all over the world, cameras with an eye on the public are increasing.

"Why just let the military and law enforcement play with this stuff?" he asks. "You can use a flame-thrower for art. You can play with surveillance art, too. Technology isn't bad, it's what you do with it."

These days, Spelletich is playing with a 12-foot-tall machine that reads brain waves and responds with a hug, and a robot hooked up to leeches.

"I'm a bit of a political animal," Spelletich says, as if that weren't obvious. "I find the leech to be a great metaphor for a lot of the problems with the planet and our society." —*Megan Mansell Williams*

Kalopticon in Action: makezine.com/go/kalopticon
Kal Spelletich: seemen.org/kal.html

Photograph by Kal Spelletich

Ship of Bricks

When **Malle Hawking** watched a documentary about the *USS Harry S. Truman*, something clicked. He'd built ships out of Lego as a child, so it occurred to him that he'd like to have a model of the American supercarrier. He found his old boxes of Lego in his basement in Munich, Germany, and started building.

After 14 months and countless shipments of parts, his creation was done. "I can't think of anything else in my life that was so exhausting and challenging," says Hawking, 38. "And for the costs, I could have had my own Volkswagen instead. But it wouldn't give me that much fun."

The ship's 54 aircraft alone took a whole month to create. There are working lights, movable elevators and radar dishes, even a webcam mounted inside the hull to let viewers see the interior.

Getting the shape of the *Truman's* superstructure right proved to be the biggest challenge, as Hawking had only one video and a couple dozen images from the internet.

More than 300,000 bricks were needed to complete the model, with its hull of light gray pieces, and internal, nonvisible elements made up of random colors. The ship is 16 feet long and weighs more than 350 pounds. It can be split into six sections for (relative) ease of transportation. When completed in 2006, it set the world record for largest Lego boat. (It's big!)

The biggest surprise for Hawking was the enthusiasm that the *Truman* met with as he began to show it off. It was blogged more than 200 times before he even finished the model. It's been displayed in Munich, Cologne, Berlin, and the Lego Museum in Billund, Denmark, and this year it appeared in shopping malls around Germany.

Even the U.S. Navy has shown interest. Recently, a deal to ship the model to the United States for display fell through, but Hawking hopes it will happen soon.

—*John Baichtal*

≫ **Lego Aircraft Carrier:** makezine.com/go/legoship

Photograph courtesy of Malle Hawking

Making Things Work

The most incredible headline of the year, if not the decade, or century, just passed through the news wires, largely unnoticed. We, humanity, built the Large Hadron Collider (LHC) at CERN, in Switzerland and France. We turned it on. It worked.

Why is this the coolest headline of the decade? It took more than 8,000 physicists from 85 countries to design it and put it together — countless engineers and technicians, with countless opportunities for a mistake to be made, or a number to be miscalculated. It's a stunning testament to the power of humans to work together to do something incredibly complicated. And it *worked*.

When I visited early in 2008, it felt like going to the shrine of technology. It is a machine like no other, a 27km loop of engineered beauty. The magnets and sensors fill underground caverns the size of buildings. It's a magical world of copper, silicon, aluminum, and stainless steel.

Why does it impress me so, this feat of engineering? Probably the first thing that most of us independently make is a finger painting, or a stack of wooden blocks we imagine as a castle. It's where we start in this amazing, lifelong process of learning to build complex things from simple parts. Steadily, we learn through life to imagine new and ever more complex things, and how we might make or build them. The process of building something new and complex from nothing but an idea is wonderful to me.

I like watching this process at all different scales, and I see unbelievable reward on the faces of people who successfully turn their concepts into things that work. I know from my own experience that watching something that you've conceived actually work as planned is the drug that keeps makers, scientists, engineers, and artists addicted to what they do.

As we grow in skill and expertise, we learn to finish things, and to finish them with craftsmanship and higher quality. As we grow in knowledge about the physics of the universe and how materials work, we learn to do a very beautiful thing: we learn that by building models — blueprints, equations on paper, computer simulations, and the like — we can make

something never made before. And we learn that the mastery of two skills — predicting the behavior of machines, and knowing how to build them to the right specifications — makes this drug particularly powerful.

It is testament to the power of our little ape brains, that we are able not just to control, but to predict ... and to be right.

We also start learning to integrate other people's sub-components and thoughts and ideas into a product or project. People specialize, and then engineering teams can be built — things that collectively can build very complex items that are truly astounding when you sit back and look at them.

A car, any car, represents an amazing amount of engineering, and the fact that the car will start as it comes off the production line and be a working machine is quite incredible. An MP3 player or new cellphone is equally amazing in its design and engineering implementation.

As individuals who contribute to the long list of science and engineering and design achievements that go into any complex thing, we makers find a wondrous joy and pride and fulfillment in contributing part of the puzzle.

And the Large Hadron Collider is the ultimate wondrous expression of all of those skills and desires to make something new and beautiful that works. It is a Swiss watch of unprecedented precision and scale. Oh, how I could wax lyrical. But the story is even better than everything I've written so far.

Why did we build the LHC at all? It is, in fact, just a science experiment. But what an experiment! Humans imagined, then designed, and have now built a machine conceived to answer the deepest questions we know how to ask about the physical universe around us.

We know that things have mass; they are heavy to lift. But we actually don't know why the elementary particles that comprise matter have mass. This incredible machine should be able to tell us why.

After a few days of operation, a faulty connection between two magnets triggered a shutdown of the LHC. It will mean a delay (the LHC will be operational again in spring 2009, after a winter dormancy), but

Watching that thing that you have conceived actually work as planned is the drug that keeps makers, scientists, engineers, and artists addicted to what they do.

MASSIVE MARVEL: (clockwise from top) The heaviest piece of the Compact Muon Solenoid particle detector being lowered into place; in order for technicians to get around the 27-kilometer tunnel that houses the LHC, various methods of transportation must be employed; when this photo was taken, the main barrel of the Atlas detector was yet to be installed, giving impressive views of the eight torodial magnets.

if that were the biggest problem I had turning on a machine like that, I'd be stoked.

Apart from just impressing me with wonder, the LHC project makes me feel good about being human. That we can imagine, theorize, model, and understand our universe. That we support science and knowledge. That multinational teams of exceptional individuals can work harmoniously to make something profound.

It gives me hope that we can actually solve humanity's larger challenges: water, energy, and sustainability.

Make something beautiful. Make it work.

Saul Griffith is a co-author of *Howtoons*, a MacArthur fellow, and CEO of a wind-energy startup.

The First Picture Show

Jack Judson reveals the beginnings of the entertainment industry at his Magic Lantern Castle Museum. By Dale Dougherty

The magic lantern was the earliest example of how projection could be used to tell stories. From as early as the 1500s, the large-scale images projected by the magic lantern fascinated people and could make them laugh or cringe while sitting together in a dark room. The magic lantern is the first technology of the entertainment industry, as well as the predecessor to the more mundane PowerPoint presentation.

Collecting magic lanterns became an obsession for Jack Judson late in life. After retiring from business he began collecting hundreds of magic lanterns, mainly from the 19th century. He not only had to learn how to build and organize his wonderful collection of magic lanterns and research their history; he had to learn how to create his own museum to preserve the treasures he's gathered.

The Magic Lantern Castle Museum is hidden away in a stretch of strip malls in San Antonio, Texas. The only flourish on the exterior of its nondescript building, which was once a disco, are the castellations. Inside, Judson has transformed the space into a private den where he can share his remarkable collection, alongside a workshop where he keeps the magic lanterns in working condition.

I took a tour of the museum and then sat down with Judson to explore the rich history of this fascinating technology.

Dale Dougherty: The first thing we see in your museum is a statue of a lanternist.
Jack Judson: The lanternist was a traveling show-man who carried a magic lantern on his back, and out front he's carrying a hurdy-gurdy, and he would

Photography by Michael Thad Carter

be walking, probably, into a town square somewhere in France, where we presume a lot of it began.

DD: What is a hurdy-gurdy?
JJ: The hurdy-gurdy is really a violin in a box, but instead of being stroked by a bow, it is stroked by a wheel that is turned with an outside crank. On the front of the box is a set of keys that can change the pitch and tone, to play music with it. He would play the hurdy-gurdy to attract attention. He might be invited into a home, or a church, where he would slide pictures painted on glass through this machine that was lit by a little oil lamp. It was basically a tin box with a lens at the front.

DD: The lanternist was essentially a storyteller who had images to accompany his story.
JJ: The magic lantern was very scary to people who had no education whatsoever. Frequently, they did these shows in total darkness, like in a crypt, which was very spooky, and they showed pictures of skeletons and devils. It just scared everybody like crazy. They might also do the projection from behind the screen. They would have a light-colored cloth, which they wet to make it more translucent.

The image could be small, or grow larger, and this was quite alarming to some people. They thought it was magic.

DD: The light of an oil lamp flickers. And it's a yellowish light.
JJ: It's a terrible light. The next evolution, of course, was trying to improve the light — the amount of light — by adding, instead of one little wick, a bigger wick or two wicks, or three wicks, or four wicks. They were able to grind better lenses. Then, of course, they were using fire — that was about as far as you could go with burning oil or some kind of a liquid, burning agent.

DD: Some lanterns used a pair of lenses.
JJ: They had a condenser lens, which is right at the front of the box with the light in it. That was either a single or double plano-convex large lens, which acted to focus the light from the source into a coherent path. It would pass directly through the image area on the slide, and then meet the projection lens out front, which you use to focus the image. That's the normal configuration. That exists even to this day in the latest slide projector.

Maker

CLASSIC TALE: The Ratcatcher, the most popular of all mechanical slides, portrays a sleeping man in bed, with a rat coming from under the covers to see the source of snoring, then crawling into the mouth. One piece of glass levers the mouth open and closed, and the crank rotates a piece of glass to move the rat up and into the mouth. Note the round brass rack and pinion operated by the crank.

DD: Let's talk about the slides. The slides are made of glass inside a wooden frame.

JJ: Exactly. A lot of them in the early days were simply freehand paintings on glass. They're miniature paintings, but, of course, made to blow up to incredible sizes at times. Some of them would be 3 inches in diameter, some of them even smaller. It was the earliest AV.

DD: That lanternist was the AV man.

JJ: He was it!

DD: Then the slides begin to change because of photography.

JJ: In the late 1830s to 1840, we got photography but no one was thinking of projection. They were making pictures on metal or paper. Fortunately, a pair of brothers from Germany, William and Frederick Langenheim, figured this out. William fought in the Texas Revolution, and Frederick started a photography business in Philadelphia. They are credited with inventing the first black and white photographic lantern slide. They didn't have color photography.

DD: You showed me examples. The process was to paint a larger picture and take a photograph of it, and then do this transfer process to create a slide, which was then handpainted to add color.

JJ: To handpaint all the details required incredible skill and eyesight, and a lot of technique. They would take characters out of, say, Tennyson's poems, or *Les Misérables*, and show the characters in great detail, because they went with published stories that were well known. They could then bring a story to light.

DD: A lot of the language of film editing originates with the magic lantern.

JJ: The first motion of any kind, or any effect, that we now take for granted — whether it's electronic, or on motion picture film, or digital — was done when they learned that you can move one piece of glass past another piece of glass, and cause things to darken out, or to change. It gives the simulation of motion.

They also learned that they could dissolve — a word we use today — from one image to another by raising the firelight in one lantern, and lowering it in the other one using a very-nearly identical slide. A house might be shown in daylight, and dissolve into an image of the house at night.

DD: Did it require dual projectors?

JJ: You would, generally, have at least two sets of lenses. That way you could dissolve smoothly without interruption of the viewing.

DD: What's your favorite projection?

JJ: The Ratcatcher slide is legendary. Basically, it was the hit of the show and I use it still. There's a man recumbent in a big old bed in the 1800s, and he's got a candle burning on his nightstand, and he's under the covers. He's got a long black beard and wears a nightcap. One of the levers on the side of the magic lantern moves a piece of glass up and down, so that his jaw opens and closes as if he was snoring.

Then you have a crank on the other side and as you turn it, coming up from under the bed and up over the covers comes a rat to investigate the man snoring. It gets closer and closer to the man's mouth. Finally, he swallows the rat. The audience goes nuts.

DD: In your museum are handbills used to promote magic lantern shows. The programs were not just stories, but also lectures — travels in England, for example.
JJ: Yes, and there are many on the evils of drinking. That was a big movement in England, called the Band of Hope, and their motto was, "Water is best." A very popular thing was catastrophes — the Youngstown flood, and the Galveston hurricane, and the terrible fire somewhere, not to mention the San Francisco earthquake. As one fellow wrote in his autobiography, people seemed to love to go see horrible stories.

DD: A magic lantern show is a group of people sitting in a room, watching "horrible" images on a wall.
JJ: They also did science lectures. Some of the shows were humorous. Some of them were educational. They used magic lanterns in churches to project hymns.

DD: One focus of your collection is how the secret societies used the magic lantern for initiation ceremonies and to reveal secrets that only the members knew.
JJ: Masons, for instance. They came up with a marvelous device known as the hoodwink. Those to be initiated were fitted with what looked like a set of goggles attached to a leather hood. The goggles had a lever on either side where you could flip open the eyepieces to see, or close them to keep the initiate in the dark. They strapped it around the initiate's head, and led him into the inner chambers, where he was shown a light-show presentation that told the secret story of the lodge. That device gave rise to the term "being hoodwinked."

DD: The magic lantern comes to be part of the early film industry starting in the late 1800s. The Edison kinetoscope could project from slides and film.
JJ: You had Edison's home kinetoscope, and, of course, then the projecting kinetoscope, which was the one that was used by more professional people.

"Man has been fascinated by projected imagery ever since there were shadows dancing on the walls of a cave."

You could buy slides for 50 cents apiece. You could not buy films; you had to rent them. Netflix of the day, I guess you might say. There's nothing new.

DD: Those early films, though, were not very long, were they?
JJ: No, they were very, very short. The earliest ones were 50 feet, which is basically the length of the table where George Eastman could lay out the film — a liquid — and let it solidify, and then roll-cut strips that were 35 millimeters [wide], and so at 16 frames per second, it doesn't last very long.

At some point, I recall the story where this old man talked to Edison about how to show these films, and he said, "Well, just run them through three times so that they get their money's worth."

There was no story. They had no message — no nothing. They were just images of people moving, and, in fact, they were not moving. They were really sequential stills. Films for the Edison home kinetoscope were printed in three tracks on one film width so the film could be run forward, then played again reversing the reel, and then again forward. It was a very unusual thing.

DD: These are hand-cranked machines.
JJ: They're all hand-cranked. It's a wonderful, clicking, mechanical sound that we don't hear anymore.

DD: You have a beautiful collection here.
JJ: Nowhere else in the world can you go and see the complete variations on how magic lanterns were made and what they were used for. It really was AV in every sense of the word, and it developed into motion pictures. Man has been fascinated by projected imagery ever since there were shadows dancing on the walls of a cave.

The Magic Lantern Castle Museum follows the use of magic lanterns up until the first generation of film projectors. Judson decided to stop collecting there, at the advent of cinema.

PICTURE SHOW: (top) The Lodge Lantern Vertical Rotary Carousel; (bottom) various 20th-century coming attraction slides. *Wings* (bottom row, third from left) was the first winner of the Academy Award for Best Picture; (opposite) the children's magic lantern exhibit includes 19th-century handpainted slides in wood frames.

"Walt Disney began working for the Kansas City Slide Company as an illustrator for magic lantern slides," Judson adds. "That's how he got his start. Of course, everybody knows how he went on from there to California where he created cels that would become an animated motion picture. All of this stuff began with the magic lantern."

» Magic Lantern Castle Museum: magiclanterns.org

A complete video and transcript of this interview are available at makezine.com/16/lantern.

Dale Dougherty is editor and publisher of MAKE.

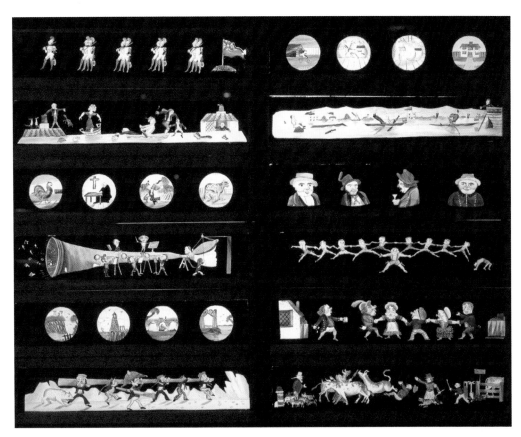

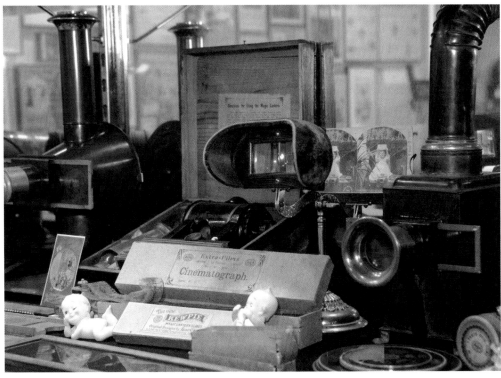

CANDID CAMERAS: Two handheld HD cameras close in on Maker Workshop host John Park.

24 Hours of *Make: Television*

Building a TV show is a project in itself. By Dale Dougherty

t's already late on a Sunday in September in St. Paul, Minn. In the studios of Twin Cities Public Television (TPT), there are ten people working on *Make:* television, a new PBS show that will be a companion to this magazine.

The set is a workshop, within a larger workshop normally used for set construction. The team is shooting a build for Maker Workshop, the segment that shows viewers how to make something in each episode. It's important to get this segment right. It's like demonstrating a recipe on a cooking show, but the ingredients and the process are more technical.

On the edge of the set are two plasma screens, one for each of the handheld HD cameras. One of them displays a timecode: 15:40:31:01 and

running. The process of making video is all about managing time.

Get Me Rewrite

Bill Gurstelle is weary, and I can tell he just wants the day to end. A contributing editor to MAKE, Bill is the technical consultant for the show. He's taken projects from the magazine and designed builds that can be demonstrated in the allotted time of just seven minutes.

Bill grabs a box of parts and places it on the workbench. It's the Pole Cam project featured on page 108 of this volume, a close cousin of our very first magazine project, Kite Aerial Photography. With this rig atop a tall pole, you can capture unusual

Photography by Matt Blum

perspectives. What's new is that the rig uses a radio-control transmitter/receiver and two servomotors to control the position of the camera and snap the pictures remotely.

Richard Hudson is in charge on the set as the show's executive producer. Knowing he has the crew until 7:00 tonight, he wants to get a few pages into the script for this project and then finish it tomorrow. He's providing the momentum, but things move slowly. He grabs the script and looks at Bill. "We have to explain servomotors without a lot of jargon," he says.

Bill thinks for a second and says: "When you turn on a regular motor, it runs. A servomotor moves a specific distance." Bill turns the switch, causing the servo to move.

The script, which Bill wrote, is now labeled revision 11. To Bill, the changes seem endless, and needless. To Richard, they are a series of ever more precise refinements that aim to use as few words as possible to accompany a series of actions demonstrated in front of the camera.

"That's perfect," Richard says.

Here's Johnny

"Where's John?" someone on the set asks, and another person answers mockingly: "He's in his trailer." He's actually in a small room nearby changing his shirt.

John Park is the host of the Maker Workshop and at 17:56 he walks in, ready to go. The Pole Cam is his second workshop segment of the day. Earlier, he built the Burrito Blaster, a variation of the potato cannon featured in MAKE, Volume 03. John, who works in Burbank, Calif., at Walt Disney Animation Studios, came in on Friday night. All day Saturday was spent rehearsing the four projects they will shoot Sunday through Tuesday.

Six people huddle around John and they talk about the sequence of the build. Richard brings up the idea of explaining servomotors. "Oh," John replies, "a servomotor has a feedback loop using pulse width modulation …"

Richard interrupts him and the entire group starts laughing. "Simpler," says Richard.

Bill chimes in with his definition and John tries it out in his own words. "A regular motor spins when you turn it on; a servomotor moves a certain distance." He practices another line: "On our rig, the servos allow us remotely to tilt the camera up and down, as well as push the shutter button down."

The script has about 18 separate scenes for this

The Maker Workshop is like demonstrating a recipe on a cooking show, but the ingredients and the process are more technical.

build. The goal is to get one or two scenes done before breaking for the night. Once the lights are arranged on the set, Greg Stiever, the director, places the two cameramen. Camera A is the focus for John when he speaks, while Camera B closes in on what John is doing. Vern Norwood, the sound guy, asks John to count to ten to test his mic. John gets to four when Vern interrupts him: "Brilliant. Most people don't get that far."

The first scene has John introducing the project. He starts off with a yellow Mr. Longarm extension pole in hand, then he'll move to a workbench to introduce the rig and the servomotors. He rehearses the scene once but Richard doesn't like something. "There's so little to look at. Put him on a stool next to the grinder."

"Grab your pole, John," says the director, getting everyone in position for the first take. "Action."

"That's awkward," says Michael Smith, the series producer, watching the scene on the plasma. He suggests a different way for John to hold the pole so it doesn't cross between him and the camera. They start again.

This time John gets further but he's stopped short again. "How is he supposed to be holding the motor?" Richard asks.

18:41 and there's a loud crash of glass. In another part of the room, a fluorescent light tube fell from a 20-foot ceiling — inexplicably. The crew takes note, but they keep things moving. "Action."

There are five consecutive takes. Each time, John amazingly dials in the same energy level and focus, making any changes asked of him, and seldom introducing anything new or different that might not be wanted. It's a lot harder than it seems. John's tired but it doesn't show.

"Mark it. That's good," says Greg after one more take, but then he adds: "Let's do it once more."

At 18:50, Richard says, "Wrap. We're done." The next day we're going to the zoo.

Maker

MAKE TAKES: (clockwise from top left) Executive producer Richard Hudson and John Park talk about the Pole Cam rig. Director Greg Stiever reviews the script while Park rehearses his lines. Series producer Michael Smith inspects the assembled rig. The crew discusses the build sequence for the Pole Cam.

Bird's-Eye Hitchcock Moment

It's 8:45 on Monday morning outside the Como Park Zoo and Conservatory. There's a water garden with lily pads and an incredible plant called a Victoria water platter that's 3 or 4 feet across. We're here to show the Pole Cam in action at the zoo, and we've come before the day's visitors arrive.

In the first scene, John will stand in a grassy courtyard and say, "Nothing beats a pole-mounted camera," then he'll turn around and say, "Hi, I'm John Park." The cameras are shooting him from the top of a two-story building. Time and again, he does it.

The next scene shows the Pole Cam in action, and John can't do it alone. So I'm holding the controller while a production assistant holds the pole. Richard says it's my "Hitchcock moment" where I get to appear inconspicuously in the shot.

We wrap up at the zoo and go to the park next door to shoot a scene for a different project, the $14 makeshift Steadicam that appeared in the first issue of MAKE. It's a tripod with a barbell weight used as a stabilizer. "I don't know if the five-pounder is good enough," worries one of the production assistants.

"We'll just have to try it," Richard says. An Ultimate Frisbee player shows up and soon three people are tossing the Frisbee. John steadies an HD camera with the stabilizer and he's running back and forth after the Frisbee. A cameraman follows John. Unfortunately, this part of the park is inhabited by Canada geese, and their droppings remain underfoot. At 10:27, the scene is done but everyone needs to clean off the bottoms of their shoes.

Drill, Baby, Drill!

At 11:16, we're back in the workshop. Bill and Richard are reviewing the build sequence for the Pole Cam's two-piece wooden rig. There's an upper frame that must fit inside a lower frame. The camera and one servo are attached to the upper frame, and another servo is attached to the lower.

Bill reminds Richard that he had to trim a tab off the servomotor with a knife. "Do we have to show that?" he asks.

"No," replies Richard. "We'll put it in the written instructions."

At 11:46, John arrives on the set. At 12:46, with several pieces of wood on the table, John begins the scene: "First, we'll build the rig." There are lots of starts and stops. A battery runs out on the wireless mic, causing a restart. The scene ends with John saying he's ready to drill a hole in the frame. "Do you want to go to the drill press next?"

There's some debate about what to do next, but it's then decided: "Let's drill." Soon a chorus of "Drill, baby, drill!" rings out, repeating Rudy Giuliani's infamous line from the Republican convention, which was held in St. Paul two weeks earlier.

Sneak Peek

I pop into the editing room to review the Maker-to-Maker segments featuring Mister Jalopy.

A contributing editor for MAKE, Mister Jalopy gets a chance to show off his garage and talk about what he discovers from "garage saleing" in Los Angeles. He talks about a vintage car he bought for "a fistful of dollars and an old bike," and why he won't restore it. It's great stuff.

At 15:06, John is tightening a nylon wing nut to join the two frames. "Now we can test out the pivot," he says. The servomotor moves the upper frame, and John smiles when it works out. Michael says this scene is the longest, covering 45 seconds to a minute. We've done about 12 scenes this afternoon, each requiring four or five takes.

In the final scene, John is supposedly looking at the pictures on the camera that we took at the zoo. "Excellent," he says looking at the still camera. "Awesome. Fantastic. Incredible. Woweee." He keeps riffing until everyone is laughing.

"Just say 'excellent,'" adds Richard.

At 17:43, we're done for the day. John's been "on" for most of it, a kind of marathon. It's about 24 hours in real time, 12 hours in actual recording time — all of it for seven minutes of a half-hour show.

I ask Bill how long it might take a person to do the Pole Cam project and he says: "About two hours if you have everything ready to go, but it would probably take most people a full day."

A full day. So, the making of a Maker Workshop segment becomes a project in itself. Not surprisingly, it's a group of people working together on deadline.

Dale Dougherty is editor and publisher of MAKE and CRAFT magazines.

***Make:* television is coming to public TV in January 2009 —** contact your local station for airtimes, and visit makezine.tv to learn more.

Based on an idea by MAKE editor at large David Pescovitz and myself, *Make:* television is a blend of "meet the makers" documentary with a hands-on workshop that shows viewers how to build things themselves. The show is comprised of four segments:

Maker Profile: A documentary segment that shows the creative and collaborative side of making. We visit San Francisco's Cyclecide group that makes human-powered carnival rides; author Syuzi Pakhchyan from Los Angeles, who designs electronics into clothing; and many other amazing makers.

Maker Workshop: Your host John Park shows you step-by-step how to make a VCR Cat Feeder, a Burrito Blaster, a Digital TV Antenna, and many other projects.

Maker-to-Maker: Insights and tips from notable makers, including Mister Jalopy, Cy Tymony, and Bill Gurstelle.

Maker Channel: Videos created by makers themselves. If you have a video you'd like us to consider, tell us about it at makerchannel.org.

➕ Read bios of the *Make:* television team at makezine.com/16/maketv.

Geek Squad: Take the World Apart

The Geek Squad was quick to sign onto the project as a major sponsor. Their founder and CEO, Robert Stephens, explains, "When I was a kid, my parents let me take things apart, and that gave me a curiosity for how the world works.

"This is why *Make:* television is important," he adds. "We need young people to be curious and take the world apart to see how it works. From Wikipedia and YouTube to MAKE, the world has an edit button on it now. The Geek Squad is proud to be a founding sponsor of *Make:* television." We at MAKE agree!

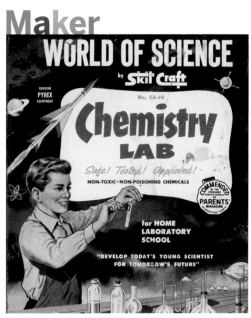

Photography courtesy of Chemical Heritage Foundation (top left, bottom left); by Dustin Fenstermacher (top right, bottom right)

BACK IN THE DAY: (clockwise) Skil-Craft sent imaginations into space in the 1950s; kids in 1965 still got real chemistry sets with burners and balances; Gilbert's 1958 girls' set was a nice gesture but had no chemicals; Lionel-Porter Chemcraft's beautiful sets ruled the 1960s.

Great Balls of Fire!

Why old chemistry sets were better — and how to make your own today.
By Keith Hammond

It's true: chemistry sets today don't measure up to the classic kits that once scorched Formica kitchen tables across the nation. But you can still find respectable kits if you know where to look. More importantly, anyone can make their own flaming, fuming, booming DIY chemistry set as good as those from the golden age — or better.

Danger Is My Middle Name

How good were the old sets? They were certainly more exciting, stocked with iodine and nitrates good for making unstable explosives or homemade rocket motors. Chlorine and cyanide compounds could emit deadly gases. A few chemicals turned out to cause cancer.

Kits from the 1920s to the 60s might include radioactive uranium, deadly sodium cyanide, or pure magnesium foil that burns at 4,000°F, with manuals that told how to mix up gunpowder or melt sand red-hot to blow your own glass test tubes. *The Golden Book of Chemistry Experiments* debuted in 1960, packed with risky experiments. Its 19th-century predecessor, *The Boy's Own Book*, had 20-plus pages of chemistry and fireworks recipes.

People tolerated more risk back then, but in exchange, generations of young experimenters were rewarded with deeper discoveries, bigger thrills, and the satisfaction of daring to achieve something important for the future.

Rocketry, nuclear energy, plastics — new sciences that were changing the world — were all highlighted in popular chemistry sets of the mid-20th century. Many of today's scientists and engineers trace their careers back to the excitement of that first set.

Kits Today: Wimpified

Compared to their robust ancestors, chemistry sets today are wimpy. They revolve around low-energy reactions and the quiet creation of crystals and polymers. The average set from the mall has no burner to provide a flame, no chemicals that go bang. It'll let you prepare solutions that change colors or glow like a light stick, but that's about it for excitement.

Why? It's common sense to delete highly toxic compounds, and we're certainly more focused these days on insulating kids from risk.

But mostly it's fear: of liability, of terrorists, of the neighbors. Overreacting to methamphetamine trafficking, Texas has outlawed the Ehrlenmeyer flask. In August, panicky Massachusetts police ransacked the basement lab of retired chemist Victor Deeb, who was simply fiddling with experiments in his home.

A Few Good Kits

But not every kit maker has chickened out. Thames and Kosmos of Portsmouth, R.I., sells the Chem C3000, a tolerably well-stocked set with extra bottles for risky stuff like hydrochloric acid and sodium hydroxide, which you're encouraged to purchase separately.

But if you really want to do chemistry at home, you'll want to make your own DIY chemistry set. Elemental Scientific sells kits of chemicals, glassware, and lab equipment selected specifically to accompany MAKE author Robert Bruce Thompson's *Illustrated Guide to Home Chemistry Experiments*. To learn more, get the book at makershed.com or visit homechemlab.com.

Keith Hammond is copy chief of MAKE and CRAFT magazines. He fondly recalls reading *Scientific American* and experimenting with his first chemistry set as a kid in the 1970s.

DIY CHEMISTRY: Then and Now

We asked author Robert Bruce Thompson about the powerful stuff in classic chemistry sets that's missing today — and where you can get it. (You can see more photos at makezine.com/16/chemsets).

Mr. Wizard's Experiments in Chemistry, Set MW-073

Owens-Illinois, Inc., Toledo, Ohio, 1973
"Exciting and Fun"

Iodine, I_2

"Iodine is now a Drug Enforcement Administration List I material, which means it's no longer readily available, and paperwork is required," says Thompson. The only exception is for 1 fluid ounce or less of iodine solution that contains 2.2% or less of iodine.
» DIY: "You can make your own iodine crystals from potassium iodide (KI), which is the subject of our first how-to video at homechemlab.com."

"Mystery Powder"

This was reportedly sucrose and acetylsalicylic acid, i.e. sugar and aspirin.

2,4-Dichlorophenol, $C_6H_4Cl_2O$

A toxic ingredient in herbicides and pesticides, it's a suspected carcinogen and endocrine disruptor. "A common precursor for industrial-scale syntheses, it's a chemical with few or no uses in a home lab," Thompson says.
» DIY: It's a mystery.

Photography courtesy of Chemical Heritage Foundation

Maker

The BGL Chemical Set

B.G.L. Limited, London, England, 1930s
"Perfectly Harmless!"

Lead acetate aka sugar of lead, $Pb(C_2H_3O_2)_2$
A highly toxic compound used as a sweetener in ancient Rome, it's hard to believe this was in kids' sets, but here it is — sold by the British Gas Light Company.
» **DIY:** Textile, dye, and alternative photography suppliers.

Sodium hydroxide aka caustic soda or lye, $NaOH$
Not really kids' stuff either. Extremely corrosive, it burns skin on contact. It reacts with acids violently, with metals to produce flammable hydrogen gas, and with sugars to form deadly carbon monoxide gas.
» **DIY:** Sold at hardware stores as crystal drain cleaner (check the label to make sure it's pure).

Chemcraft Chemical Outfit No. 1

Porter Chemical Co., Hagerstown, Md., 1917
"Perfectly Safe. Contains No Poisonous Or Otherwise Harmful Substances"

Potassium nitrate, KNO_3
Sulfur, S_8
Gunpowder precursors were common in chemistry sets before about 1940. Black powder is potassium nitrate (saltpeter), carbon (charcoal), and sulfur mixed in the correct proportions (*see MAKE, Volume 13, "The Fire Drug"*).

"Potassium nitrate and table sugar, if processed and mixed properly, can form a low explosive that's used for amateur rocket motors ("rocket candy")," says Thompson. "Otherwise they simply form a very combustible mixture that burns fiercely and generates a lot of smoke (smoke bombs). In today's sets, sulfur is still common — but not potassium nitrate."
» **DIY:** Charcoal you can get anywhere. Potassium nitrate is sold as a fertilizer (14-0-45 or 13-0-46), and sulfur is sold in lawn and garden stores to control plant pests and diseases.

Ammonium nitrate, NH_4NO_3
Explosive, it's used in fertilizer bombs like the one that destroyed the federal building in Oklahoma City in 1995.
» **DIY:** Sold at lawn and garden or farm supply stores as 34-0-0 fertilizer.

Strontium nitrate, $Sr(NO_3)_2$
Extremely volatile, it's used to color fireworks. When heated, it releases toxic nitrogen dioxide gas.
» **DIY:** Chemical supply companies.

Photography courtesy of Chemical Heritage Foundation; by Dustin Fenstermacher (second from top)

Gilbert Chemistry Set

A.C. Gilbert Co., New Haven, Conn., 1920s
"Today's Adventures in Science Will Create Tomorrow's America"

Sodium cyanide, NaCN

Erector Set inventor A.C. Gilbert actually sold kids this chemistry set with sodium cyanide, the stuff of suicide capsules and murder most foul. "Deadly stuff in pretty small doses," says Thompson, "just like potassium cyanide. It also reacts with acids to form hydrogen cyanide gas, which is also deadly. This isn't something kids should be messing with. It's so toxic that it'd have been insane to include it, even back when things were a lot more relaxed."
» **DIY:** It's sold by some photography suppliers, but its transport is heavily regulated.

Gilbert No. U-238 Atomic Energy Lab

A.C. Gilbert Co., New Haven, Conn., 1950s
"Most Modern Scientific Set Ever Created!"

Radioactive uranium ores, UO_2 and others

These 4 small samples of carnotite, autunite, torbernite, and uraninite emitted alpha, beta, and gamma radiation. The set also had a Geiger counter, a cloud chamber to see the paths of alpha particles, and an electroscope and spinthariscope for detecting radioactivity and decay.
» **DIY:** "Readily available," says Thompson. "United Nuclear sells small chunks of various (slightly) radioactive ores and minerals. They present no real danger, although they shouldn't be ingested and it's a good idea to handle them only with gloves and tongs."

Gilbert Chemistry Set

A.C. Gilbert Co., New Haven, Conn., c. 1920s

Glass blowing kit

A.C. Gilbert strikes again. Sand (silicon dioxide) melts at 3,100°F. But if you add soda ash (sodium carbonate) and powdered limestone (calcium carbonate), it melts into glass at just 1,600°F. Still, that's 1,600°F, kids. Mind your fingers.
» **DIY:** Readymade lab glassware is sold by suppliers like Elemental Scientific (see Resources).

Maker

Porter Chemcraft Master Chemistry Lab No. 616 featuring Atomic Energy

Porter Chemical Co., Hagerstown, Md., 1950s
"Modern Plastic Experiments. Outer Space Experiments"

Radioactive uranium ore
See Gilbert No. U-238.

Carbon tetrachloride, CCl_4
Nickel ammonium sulfate, $Ni(NH_4)_2(SO_4)_2 \cdot 6H_2O$
Both are likely carcinogens, banished from chemistry sets today. Neither is particularly dangerous to handle, says Thompson, if you take proper precautions to avoid fumes or dust, prevent skin contact, and so forth.
»DIY: Readily available from chemical suppliers such as Elemental Scientific and Home Science Tools — see Resources, below.

Calcium hypochlorite, $Ca(ClO)_2$
A strong oxidizer, it's been known to undergo self-heating and rapid decomposition, releasing toxic chlorine gas. "Frankly, I see little use for this chemical in a home lab," says Thompson. "For most purposes you can substitute the much safer sodium hypochlorite solution sold in grocery stores as chlorine bleach."
»DIY: Sold as pool and spa "shock" treatment; use bleach instead.

Sodium ferrocyanide, $Na_4Fe(CN)_6$
Ferrocyanide salts react with iron(III) (ferric) ions to produce the intense pigment Prussian blue, so they're a great test for the presence of ferric ions. "Despite the 'cyanide' in the name, these salts are relatively nontoxic and safe to handle," says Thompson. "Heating them to decomposition or treating them with a strong mineral acid does produce hydrogen cyanide gas, which is deadly in significant amounts. Technically, these salts are considered poisons, but they're not really dangerous if handled with normal precautions."
»DIY: Chemical suppliers. For most purposes, you can substitute the more readily available potassium ferrocyanide, $K_4[Fe(CN)_6]$, also available from photography darkroom suppliers.

Photography courtesy of Chemical Heritage Foundation (top); by Dustin Fenstermacher (bottom two)

RESOURCES
- » Chemical and equipment suppliers:
  Elemental Scientific elementalscientific.net
  Home Science Tools homesciencetools.com
  Science Kit sciencekit.com
  United Nuclear unitednuclear.com
  Edmund Scientific scientificsonline.com
- » Alternative sources for chemicals:
  hyperdeath.co.uk/chemicals
- » Chemical Heritage Foundation:
  makezine.com/go/chemheritage
- » Reproduction manuals for vintage Chemcraft sets: gordonspeer.com
- » Top 10 Amazing Chemistry Videos, from Wired Science: makezine.com/go/wiredscience

Junk Pedalers

A company that hauls away your trash — on bikes.
By Peter Smith

Photograph by Peter Smith

Pedal People is an 11-person cooperative bicycle business in Northampton, Mass., that hauls furniture, yard waste, and garbage — all year round. In June 2007, Pedal People signed a contract with the City of Northampton to pick up its 70 trash barrels in downtown, one of the only bicycle-powered businesses in the United States picking up municipal trash.

I spoke with Pedal People's founders, Alex Jarrett and Ruthy Woodring, from their home in Florence.

Peter Smith: When did you start hauling with bikes?

Ruthy Woodring: I started in Chicago when I was living in a Catholic Worker house. One of the things we did was an open dinner, like a soup kitchen. To get food, we would do some dumpster diving and we also had a food pickup run to get donations. One day the truck broke down and so I thought, "Well, I'll just tie the garden cart to the back of my bicycle."

I started doing that until a friend came by and said, "You know what, Ruthy? I got something better for you." And he donated a trailer to the house.

PS: Now you're doing more. What's an average day?

Alex Jarrett: Typically a day with Pedal People involves getting the trailer ready, making sure you have everything you need: bungee cords, spare trash bags, gloves, and our trailers, which we stack with eight recycling totes with lids.

RW: Our trailers can take four tubs on the bottom and four on the top. So by the time the trailer's full, it's about as high as my head. The weight limit on the trailers is between 200 and 300 pounds.

PS: Tell me about the bike trailers you build.

AJ: If you can find your own wheels and plywood, it costs about $30. Labor's about five to ten hours — if you get good at it. You need some basic tools and brazing equipment.

Go to the hardware store for a few lengths of conduit. You can use them to make a few other optional things like handles. For the dropouts, you need ³⁄₁₆" angle iron. I only have a hacksaw to cut out the dropouts, where the wheels will go.

And then the hitch is made out of ¼" flat stock and a rod you bend to go around the chain stay. You have to buy a little $6 universal joint for the hitch so the trailer can tilt as you're riding around.

PS: Do you think more people will be doing this? Is this an easy way to transport things?

RW: It's a pretty unique situation in Northampton, where the city does not provide municipal trash pickup service and does provide a transfer center within biking distance of most households. Those two factors let Pedal People work.

For other people in other towns, hauling things on bicycles is pretty easy. There's very low overhead. The bike trailers are relatively inexpensive. The capital is your human labor. Compared to a truck, you don't have to pay for gas.

» Pedal People: pedalpeople.com
» Build a bike trailer: bikecart.pedalpeople.com

Peter Smith, a freelance writer based in Portland, Maine, snacks on wild crabapples and dumpstered baguettes. His work is viewable at peterandreysmith.com.

DIEHARD RACERS: The Make: Way team members are (clockwise from top) Brett Doar, Tom Jennings, Jason Torchinsky, and Sloan Fader.

We're Number 33!

The Make: Way race team takes on the 24 Hours of LeMons.

By Jason Torchinsky

I lied to my mom about having health insurance. That is my one, solitary regret about the entire exciting, challenging experience of turning a $300 junker into a real live race car.

The reason I and my very talented Make: Way teammates, Brett Doar, Sloan Fader, and Tom Jennings, turned a $300 crapbox into a race car in the first place is so we could be a part of Jay Lamm's 24 Hours of LeMons — a racing series open to anyone with a metabolism, a modest entry fee, and a car that costs less than $500.

Car Buying on a Skateboard Budget

Finding the right car is the first real challenge of the race. Your gut instinct is to buy the fastest thing you can find, but that would be a mistake. The 24 Hours of LeMons is an endurance race, with 14 hours of driving over two days, covering about 400 miles on a challenging track, in the middle of a hot May. A car's durability under these grueling conditions is more important than its speed.

We looked for gently used, well-maintained vehicles. Most had one or more serious flaws; others were so tired they practically had a blinking KILL ME NOW light on their dashboards. What we ended up with was a boring but extremely well-maintained 1993 Ford Escort LX.

Now, the car wasn't fast — only 88 horsepower and an automatic — but it seemed capable of lasting the race, and we would do all we could to help the speed. Plus, the Escort is an extremely common car, so parts are cheap and plentiful. At $300, it was the perfect place to start.

Photograph by Amy Crilly

Performance

Converting an economy runabout into a racer is largely a process of removal. Weight is the enemy of performance, so the first step was to gut the car of anything unnecessary: carpet, seats, insulation, door panels, hatchback, glass, you name it. The less those 88 horses have to drag around, the faster we'll go.

Once the car was lightened, we started in on performance enhancements. For better weight distribution, which leads to better handling, we moved the heavy battery to the rear of the car. We removed the catalytic converter and most restrictions on the exhaust system, replacing them with a simple Cherry Bomb-type muffler. The only actual "racing" part we used was a free-flowing air cleaner, courtesy of MAKE contributing editor Mister Jalopy. As with the exhaust system, opening up the intake of a motor is a quick, cheap way to free up an extra horse or two.

Reliability

If we wanted to be the tortoise that beat the hare, we needed to stay on the track, and one of the biggest factors conspiring to get us off the track would be heat.

First we tried mounting two radiators atop the ventilated hood. This failed, and taught us a valuable lesson: on a cheap car, everything is designed to perform to spec, and not a bit more. That means the water pump can happily pump water as far as it needs to go, but God help you if you try to go an inch farther.

In the end we remounted the radiator to its original position, but added a secondary fan atop the hood and a dedicated transmission cooler made from an old A/C condenser. At the race, in 80-degree-plus heat, while many of our more sporting competitors were spewing geysers of steam, we didn't overheat once.

Safety

Safety equipment doesn't count toward the $500 limit, which is a good thing, as you don't want to cheap out on your roll cage, tires, or brakes. We were lucky enough to have the free services of a professional welder, Mike Garcia, to help us assemble our cage. There's a reasonable amount of contact in the race, so in addition to the cage to protect us, we reinforced the flimsy body, especially around the radiator. We also added a racing seat with a five-point harness, which, along with helmet and full fireproof racing gear, is a requirement.

Makers, Start Your Engines

We barely finished the car in time for the race. We had little time to practice, because the car was illegal for road use, and finding places to drive it was difficult.

The first laps of the race were both terrifying and exhilarating. Once the green flag waves, the nearly 90 other cars on the track storm like angry wasps, passing and trying to pass within inches. Every driving instinct you have is set into a panic by events you normally associate with big, expensive trouble: shrieking tires, the sickening sound of crunching metal, the smell of burning rubber — but once you realize there are no insurance numbers to exchange, you start to get into it.

There's plenty of contact in the race, and the track is quite technical — not very fast, but lots of hairpins, S-curves, and the like, so the speed disadvantage of our car wasn't so pronounced.

We were only off-track once for mechanical woes, when the front wheel flew off while I was driving, which is actually less terrifying than it sounds.

In the end, we came in 33rd out of nearly 90 cars — beating out many more obvious pieces of sporting machinery like V-8 Mustangs, BMWs, and even an Alfa Romeo. It was a far better result than I ever would have expected.

The maker crowd and the backyard, greasy-handed auto enthusiast crowd have a fairly large divide between them, which seems absurd, considering how many maker skills go into a project such as this. I encourage all makers to have a go at the dirty, loud side of making things, and the 24 Hours of LeMons is a great way to do it. See you on the track — we'll be turbocharged next time around.

✚ Make: Way, MAKE magazine's official racing team: makewayracing.com

» 24 Hours of LeMons: 24hoursoflemons.com

Jason Torchinsky (jasontorchinsky.com) is an artist and tinkerer living in Los Angeles with his partner, Sally.

Photograph by Sally Myers

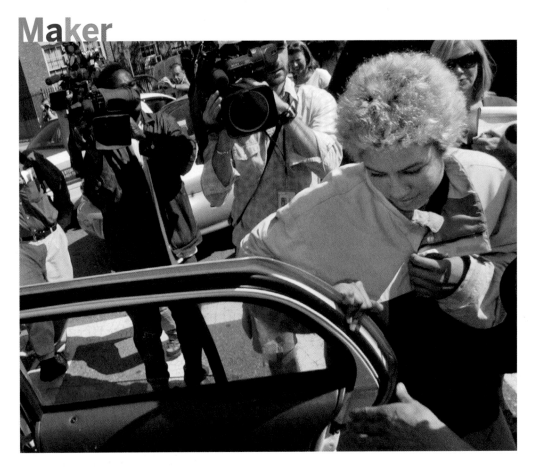

Star Bust

The "Boston Logan Terrorist Suicide Bomber" explains
what really happened to her. By Star Simpson

It's 8 p.m., and I'm nervous and excited. Nervous because I have roughly 12 hours to finish two MIT problem sets I haven't really started on. Excited because the next morning, my friend Tim Anderson will be flying from California to visit me in Massachusetts. My plan is to grab dinner, plunk myself into a study room with my books, pen, pad of graph paper, and computer, and do what's known around MIT as "tool" — that is, work solidly until I'm finished, all night if necessary. At 6:45 a.m. I'll go to the airport to meet Tim's flight.

Things go more or less as planned. At 5:30 a.m., I lay down on the solid hardwood floor of the study room to catch a catnap. I set an alarm for 6:30. It goes off, but I'm so exhausted that I doze in a half-awake state, music blaring in my headphones, until with a jolt I realize it's already 7:05! The plane is

scheduled to land in just 15 minutes.

I run down to my room and throw on my clothes in a fireman's hurry. The early morning looks gray and dewy, so I grab a sweatshirt. It happens to be the one I'd made the week before, with a battery-powered LED star on it. Tim studied electrical engineering, and he's shown me a lot about how to build things. It crosses my mind that wearing my nerdy, homespun sweatshirt might make him smile. I also pick up a little flower I sculpted from plastic that I want to give to him.

I fly outside, bleary-eyed and a little wobbly, but filled with enthusiasm to see Tim again. I rush to get to a T stop, hoping he'll wait for me at the airport.

Early-rising coffee drinkers chat boisterously all around me on the subway. After a long ride on the T, I get off at the airport stop and jump on the shuttle

Photograph by AP Images / Lisa Poole

to the terminal. I look around — no Tim. Check the baggage claim — no Tim there, either.

It's almost 7:45. I consider walking to nearby terminals, in case his plane made a gate change at the last minute. I spot an information desk. Aha! Information! Exactly what I need.

I ask the woman behind the counter if she knows anything about a change to Tim's flight.

She looks up at me and starts saying things I don't understand. She doesn't speak English as a first language, making communication difficult.

"What is that? You can't have that." She gestures at my shirt.

What? Am I supposed to give her my clothes? I start to answer her first question, about my shirt decoration. I point at it and hold it out for her.

"It's a bunch of lights, see? Decorations? I made it."

"You can't have that, what is that, you can't have that," she repeats.

"It's art, just a bunch of lights." I'm not sure how to properly explain, so I'm trying to do so in simple terms. "Can you tell me about a flight? It came in this morning from Oakland."

She's getting frantic, hysterical. For some reason, she isn't hearing what I'm saying. She appears to be completely glazed over with fear. "You can't have that. No. No, you can't have that. I'm calling the police."

I really don't understand, but I realize that nothing I say will help this woman comprehend. I'm frustrated, tired, and just really want to see Tim, so I translate "I'm calling the police" to mean "please

go away," and I turn and walk from her desk while unplugging the battery to make the lights turn off.

I make one more pass through the baggage claim and decide I must have missed Tim. Depressed, and with the weight of my problem sets to finish, I walk to the traffic island to catch the next MBTA shuttle so I can go back to school.

While I'm in baggage claim, a man dressed in black walks by me. He looks at my eyes. He looks at my sweatshirt, and continues to walk past. I watch him because he doesn't look like a passenger. The back of his shirt has "State Police" written in tall, white letters.

I think, she didn't really call the police, did she? Maybe the police make regular patrols around here.

I'm on the traffic island when someone grabs my wrists. Suddenly, shouting is coming from every direction. I feel my arms get wrenched up over my head.

People in black uniforms are all over me, yelling and forcing my arms into uncomfortable positions.

Some dam of stress breaks. I burst into tears. What's going on? I'm trying to go home and finish some crucial homework I probably won't be able to complete before it's due. I've missed my friend at the airport because I overslept. And now I'm getting mobbed by a gang wielding guns. It seems like 40 people are surrounding me. Some of them are holding pieces of metal that I initially mistake for giant camera tripods. They turn out to be German MP5 submachine guns.

"I'm an MIT student!" I shout.

"Empty your pockets! *Slowly!*" they shout back.

"Does she have a lighter on her?" one sergeant shouts to another, hoping, I later realize, to bolster their "hoax device" argument with the idea that the flower in my hand could be a blob of plastic explosive.

For the next hour, I'm an "alleged" MIT student, until someone at my school can confirm it. Which MIT does, by issuing a press statement disowning me, based on the trickle of lies fed by early news reports.

"As reported to us by authorities, Ms. Simpson's actions were reckless and understandably created alarm at the airport," reports my school's news office. Their statement doesn't help me feel any better.

"What is this?" an officer shouts, holding up the metal U-lock clipped to my bag.

"That's a bike lock," I respond.

"Why would you bring a *bike lock to an airport?*"

In the post-9/11 era, everything is suspect. I supply a reason for carrying a bike lock: "I bike."

Within a few minutes of surrounding me and demanding I remove my sweatshirt for their inspection, the police and bomb squad realize my sweatshirt is innocuous. Nevertheless, they arrest me, put me in handcuffs, and take me to the State Police office for more questioning. I'm "processed" and sent to a room where a detective questions me for an hour or more.

The telephone in the room never stops ringing. The U.S. secretary of homeland security wants to know about the "Boston Logan Terrorist Suicide Bomber."

I'm terrified, exhausted, and want to fall asleep on every remotely soft surface I see. I realize I'm not going to get my problem sets done.

Nothing makes any sense. One minute, you're an MIT student trying to be organized and do good work, and the next you feel like only Chuck Palahniuk could write a more bizarre story.

Before I know it, reporters are calling my family and everyone I've ever known.

After several hours, I'm placed in a police transport car to be taken to East Boston District Court. The radio is on, tuned to a daytime talk show. The host is trashing me by first, middle, and last names, discussing various vast, awful, and mean speculations about who I am and what happened. The officer switches it off without comment.

How does the media know my middle name, where I live, and how to reach my family? How is the radio capable of telling me what happened at the airport, before I'm even sure myself? It's evident that someone in the police department sold the story for a really nice dinner.

My house is staked out, so I stay with a friend. My old dorm is surrounded, as are my old haunts, including the machine shop at MIT. One time, I'm spit on: while I'm riding my bike around Copley Square in June — ten months after being arrested! — a man snarls, "You shoulda done time!" and hocks a giant loogie on my spokes.

For months, I can't walk down the street or use public transportation without being recognized. Many people take their cues from the same factless news reports. I can even tell which news sources a person tuned in to, by what they believe about me.

This is Boston, the same city that blew up Cartoon Network's LED signs and a private firm's traffic counter because the unidentified electronics weren't well-labeled enough to prevent the bomb squad from thinking they might be a threat.

After ten months of slow-progress court proceedings, East Boston District Court finally drops the hoax device charges. I'm ordered to perform 50 hours of community service, avoid being arrested in Massachusetts for one year, and submit an apology to the people who almost shot me because they

"You can't have that. No. No, you can't have that. I'm calling the police."

overreacted to the LEDs on my sweatshirt.

If I don't, I'll be charged with disorderly conduct, which is hard to defend, because it's not necessary for the state to prove I intended to be disorderly, only that I behaved in a disorderly way. In the end, I choose to finish the court case at the first possible opportunity, because the ordeal has exhausted me.

I'm well aware that things could have gone much worse. To quote State Police Maj. Scott Pare at the press conference, "Thankfully because she followed our instructions, she ended up in our cell instead of a morgue."

I'm disturbed by the idea that, with one hysterical phone call, the state can be set in motion to relentlessly persecute anyone. Especially in a town full of tech hobbyists. Also, the State of Massachusetts seemed unable to stop persecuting me, no matter what the facts were, once the wheels were set in motion. I don't like what this means about the future.

Of the few hilarious side effects of the arrest, the funniest involves the 2007 International Symposium on Wearable Computers. All of my wearables superheroes (especially Leah Buechley, whose LED clothing project appears in CRAFT, Volume 01) were to convene for the symposium, held, by chance, in Boston. I was invited, and I was delighted to accept.

Only there was a catch. The venue chosen was the Hyatt hotel at Logan International Airport. The judge's ban against my approaching MassPort property of course blocked my ability to attend the event. So, getting arrested both wholly created, and destroyed, that opportunity for me.

Many thanks go to my legal team of Tom Dwyer, Serina Barkley, and others at the firm Dwyer & Collora, to my parents, and to Tim Anderson, Hal Abelson, Gerry Sussman, Patrick Winston, Ken Manning, and everyone who knew better than to take the police and press accounts seriously.

Editor's Note: Star Simpson wrote a how-to for making your own light-up sweatshirt at Instructables: makezine.com/go/starshirt.

Star Simpson grew up in Hawaii and studied at MIT.

GIVE **COOL** THINGS:

Make:
technology on your time

Give your favorite geek a gift they'll love this holiday season. Send a full year of Make: (4 quarterly volumes) **for just $34.95, a savings of 42% off the cover price.**

We'll send a card announcing your gift. Your Gift recipient can add digital editions at no additional cost if they choose. Price is for US only. For Canada, add $5, for all other countries add $15 for 4 quarterly issues.

**PROMO CODE
4BDMKT**

makezine.com/gift
for faster service, subscribe online

GIVE **COOL** THINGS:

Craft:
transforming traditional crafts

Give your favorite Crafter the only magazine devoted to transforming traditional crafts. Send a gift subscription of Craft: (4 quarterly volumes) **for just $34.95, a savings of 42% off the cover price.**

We'll send a card announcing your gift. Your Gift recipient can add digital editions at no additional cost if they choose. Price is for US only. For Canada, add $5, for all other countries add $15 for 4 quarterly issues.

**PROMO CODE
4BDMKB**

craftzine.com/gift
for faster service, subscribe online

The Art of the Frame

Akio Tanabe creates some of the world's most sought-after bicycle frames. By Jess Hemerly

Tanabe-san poses for a photo in his Tokyo workshop.

Photograph by Jonathan Koshi

Shortly after World War II, the Japanese created a new form of track cycling: *keirin* (pronounced *kay-rin*). In a keirin race, a bicycle, motorbike, or moped sets pace for six to nine bike riders, gradually increasing speed on each lap. When the pacer drops off, the race becomes a sprint as riders jockey for the front position.

People across Japan trek to velodromes to watch and bet on keirin racing the way Americans bet on horse racing — except the stakes are much higher.

With significant sums of money at stake, a governing body, Nihon Jitensha Shinkōkai (NJS), regulates keirin racing. NJS has exceptionally high standards for bike geometry, weight, and materials, to ensure that a rider's equipment never provides an advantage or results in catastrophic failure. This fiercely regulated system of quality standards comes from a tradition of quality and integrity. Many keirin bikes are hand-built by a single frame builder.

Urbanites worldwide have caught keirin fever, and frames branded with some of Japan's most notable names, from 3Rensho to Watanabe to Makino, roll through city streets. And while NJS-stamped frames and parts are sought by fixed-gear fans for street riding, the NJS stamp of approval is the only thing that will allow a frame or part on the keirin track.

On a recent trip to Tokyo, we were lucky enough to hook up with a Flickr contact, a bicycle aficionado named Yohei Morita. On Christmas Day, Morita picked us up at our hotel and offered to take us around to some of his favorite bike shops in the city.

After the first shop in the Shibuya ward, we considered where to go next. I blurted out, "Kalavinka!"

"Sure, it's a short drive," our host answered.

For 35 years, Tsukumo Cycle Sports, a small community bicycle shop located in Meguro ward, Tokyo, has serviced all kinds of bicycles, from domestic *mama-chari* to professional keirin bikes.

But it's what lies in the back of the shop that makes Tsukumo a destination for bicycle aficionados. It's in this tiny workshop, the size of a large closet, where Akio Tanabe creates some of the most sought-after bicycle frames under the name Kalavinka.

There's nothing particularly new about the technology Tanabe-san uses to hand-build his frames. His workshop is filled with sketches, bottom bracket shells, lugs, and bottles of chemicals. There's no automated assembly line, no shiny new tools, and, until recently, no space-age carbon fiber. Kalavinka has been working on a carbon track frame for some time, but Tanabe-san is best known for frames produced with steel tubing and welding machines.

Before opening Tsukumo and starting his own line of bikes, Tanabe-san worked as a test rider and racer. He builds 80 to 90 frames a year, half of which are for professional keirin racers.

Despite Kalavinka's prestige, Tanabe-san is incredibly humble. He greeted us warmheartedly, showed us his workshop, and even posed for a picture. But when we began lumping on the praise, he deflected it by pulling a metal Kalavinka head badge from underneath the workbench, meticulously hand-painted by his wife.

The art of frame building is enjoying new interest in the United States. The United Bicycle Institute in Ashland, Ore., offers two-week programs for aspiring frame builders. And it's partly because of Japanese legends like Tanabe-san.

» Tsukomo Cycle Sports: kalavinka-bikes.com
» United Bicycle Institute: bikeschool.com
📷 More photos: makezine.com/16/kalavinka

Jess Hemerly is a research manager at the Institute for the Future, a freelance music critic, and a bicycle enthusiast.

TELEPRESENT: Maker Marque Cornblatt lounges at home with Sparky 2, his autonomous telepresence robot, while his cat One ignores them both.

Sparky 2: No Sellout

My robotic alter ego steps out — open source. By Marque Cornblatt

I spent much of my childhood dismantling toys and gadgets and cobbling them back together in interesting ways. One proud example combined a slot car, a one-function wireless remote, a 9-volt battery, and a few fabricated gears and bits to create (in my mind, in the early 80s) the world's smallest remote control car.

The 2-inch vehicle was top-heavy and had too much torque, but it accelerated violently to the right every time I pressed the remote button — it worked! — until it finally tore itself apart, like a tiny top-fuel dragster. In my mind it was a success, and it sparked my lifelong interest in interactive, kinetic projects.

In the early 90s, I began building a wireless, rolling, remote control robot with a two-way video chat setup positioned at eye level, which enabled real-time, face-to-face communication. I found most of the materials dumpster diving or at garage sales: a motorized wheelchair, a few old video cameras, a wireless baby monitor, and some R/C toys. Separately these were junk, but combined they became an interactive sculpture that allowed me to see, hear, chat, and move through a remote location with complete autonomy.

I could "become" my creation, temporarily merging my own identity into that of this machine/human hybrid. I named the robot SPARC-I, a rough acronym for Self-Portrait Artifact — Roving Chassis, or Sparky for short.

Sparky Works the Room

I originally made Sparky to explore the boundaries of the body and how our identities change when filtered through technology, topics that have recently become hot in our age of online profiles and avatars.

I've been upgrading Sparky ever since, as newer technologies have become available. Over the years, the Sparky experience has developed into what I call Autonomous Telepresence, an experience combining remote sensing and locomotion, web video, social networking, and human interaction.

It's interesting to watch Sparky "work the room" at an art opening or cocktail party. At first, people are drawn to the robot as a techno-spectacle. But it's remarkable how quickly people forget the machine and interact with the remote person, joking, flirting, or having long, deep conversations as if there were nothing unusual. So Sparky has informed my sense of body, self, and identity, but it has also guided me toward insights and decisions in an area I never expected.

The original Sparky had severe limitations, such as bad audio quality and a broadcasting range of just a few meters, but I improved its performance over the years by swapping in new technologies, including better audiovisual transmission components, radio control upgrades, and fresh batteries.

The biggest upgrade was the decision to leave the old-school analog AV components behind and make the leap to digital. Using the power of wi-fi and the internet, Sparky became truly remote, enabling real-time chat and control from virtually anywhere in the world.

During the mid- to late-90s tech boom in the San Francisco Bay Area, Sparky became a party machine. What tech-savvy startup wouldn't want a robot at their launch party? Sparky would mingle and schmooze for hours on a full battery charge, while I worked behind the curtain.

We made appearances at (and crashed) parties for the San Francisco Museum of Modern Art, San Jose Museum of Art, Burning Man, the E3 Media and Business Summit, Industrial Light and Magic, Intel, and others. Sparky was even, briefly, the singer and leader of a jazz quartet.

To Productize or Not?

Back then, it seemed everyone was getting VC funding for new tech businesses based purely on speculation. Meanwhile, I had created a proven, one-of-a-kind prototype that seemed to have commercial potential. Sparky's success in such varied professional and social settings inspired me to consider its potential in a wider range of environments, from facilitating distance learning to working as a museum tour guide.

I researched the nascent mobile telepresence

During the mid- to late 90s, Sparky became a party machine. What tech-savvy startup wouldn't want a robot at their launch party?

market and discovered that several Sparky-like devices were already for sale or soon to hit the market, ranging from children's toys to the $100,000-plus hospital-bots that appear on *ER*.

Given this commercial environment, I understood that it would take a lot of effort to define Sparky as unique, and then raise the venture capital to pursue designing, manufacturing, and protecting it as intellectual property.

I wrote a business plan and met with potential investors to pursue Sparky's commercial development. In the meantime, the Sparky upgrades continued. In 2006, John Celenza and I built Sparky 2 (see page 53), and I had a great time using it to telepresently cruise the exhibit floor at the first Maker Faire. Sparky 2 also appeared on the History Channel's *Modern Marvels* as a possible "future of the telephone," and I demonstrated it at technology and entertainment industry conferences.

I weighed the pros and cons of going commercial. John and I could freeze development, treat Sparky's design as a trade secret, and attempt to "productize" it under the distraction and meddling of investors. This did not sound fun or interesting to me.

On the other hand, without investment, we could keep experimenting, trying new video chat clients, motor-control schemes, and other Sparky-relevant technologies as they improved and became cheaper. We could bounce from one technology to the next, adapting what worked for our one-of-a-kind creation, and enjoying the journey free from investor expectations.

This offered a far more appealing path, and with this insight, I realized two things:

One, I'm a maker first, and a business guy second.

Two, Sparky and Autonomous Telepresence are not defined by hardware, software, or other technologies — which are changed and upgraded too frequently. Instead, Sparky is defined by the unique experience it offers, which has remained consistent over its years of evolution: an opportunity to

Maker

LEFT: Close-up of the MAKE Controller that connects Sparky's onboard computer to the servomotors. RIGHT: The exposed TV tube. Mounting the monitor without its case is a bit more dangerous, but how cyberpunk!

become a hybrid identity that can have intimate, face-to-face interactions and move freely in a remote location.

At New York's American International Toy Fair in early 2008, I saw a cheap mobile telepresence toy, clearly not designed for hackability, and that clinched it. I decided to share Sparky 2 as an open source DIY project, based on the MAKE Controller board and the components I had lying around.

DIY Sparky

So here's the take-away: you can now find a full set of plans for an open source, DIY Sparky at makezine.com/16/sparky, as well as a step-by-step video at gomistyle.com.

I co-developed the DIY version of Sparky 2 with my longtime friend and programmer extraordinaire, John Celenza. It uses the MAKE Controller connected to an onboard Mac Mini, which transmits audio, video, and motor control data over a wi-fi network.

We have several software versions, including a lag-free one that requires a web server to connect robot and controller, and another one that sends the motor-control data through Skype. We've also developed a small iPhone patch that allows it to

connect in any location, even without wi-fi. If the "Jesus phone" can connect, so can DIY Sparky.

The DIY Sparky shown here was built using scrap components I had available, including Vex kit components for the chassis and the motors, an old Mac Mini, an unused LCD monitor, an iSight webcam, a 12V scooter battery, and an AC inverter. Your version will be different, based on whatever you have lying around.

The only part I purchased was the MAKE Controller board — I could have gone with the cheaper and better-known Arduino, but the MAKE board offered additional helpful features like the 4 plug-and-play servo connectors. The MAKE board also has numerous digital and analog ins and outs, which give DIY Sparky plenty of room to grow new appendages, like movable gripper arms, sensors, and bumpers.

📷 See more of Sparky at makezine.com/16/sparky.

Marque Cornblatt (marquecornblatt.com) is a conceptual artist, roboticist, and maker. He is the creator and host of Gomi Style, a green DIY lifestyle and design series on the web (gomistyle.com).

I-Contact

One experimental Sparky design addresses a problem common to all webcam-based video chat applications: the annoying lack of eye contact.

During a video chat, we tend to look at the image of the other person on our monitor rather than into the webcam mounted above or beside it. Because of this, we seem to gaze over the shoulder of the person we're chatting with, which fails to replicate the intimacy of face-to-face communication.

The new Sparky design takes a cue from the way teleprompters work. For both Sparky's "head" and the remote operator's webcam, an angled piece of one-way glass reflects the video image of the other person directly toward viewers. Meanwhile, a video camera placed behind the glass captures the face(s) looking at the screen, creating a simulation of eye contact.

My experience has shown that this illusion of eye contact is effective. It encourages users to talk more normally, look into each other's eyes, and forget that there may be thousands of miles physically separating them.

–Marque Cornblatt

DIY-brary

Rick and Meg Prelinger couldn't find a library with what they wanted, so they made their own. By R.U. Sirius

It's an overcast Monday afternoon as I arrive at 8th and Folsom in San Francisco's seedy, bohemian SOMA district. I find 301 8th Street and then buzz Room 215. A voice says hello. I tell him who I am and he buzzes me in. I take an elevator to the second floor, walk past several closed offices, and enter a small room packed to the rafters with four rows of shelves filled with books stacked 15 feet high. This is not your typical 21st-century urbane, haute-culture library.

The Prelinger Library (prelingerlibrary.org) is the brainchild of Megan Shaw Prelinger and Rick Prelinger. Founded in 2004, it's a DIY, appropriation-friendly, intuitive, and highly personalized context for organizing and sharing this couple's books, periodicals, printed ephemera (like obscure government documents from the Department of Indian Affairs),

and — most of all — their obsessions. In addition to its physical presence in San Francisco, it has an online presence of more than 3,000 scanned volumes at the Internet Archive (archive.org).

Rick is also the founder of the Prelinger Archives, a collection of about 200,000 discrete items, including 60,000 advertising, educational, industrial, and amateur films. Rick founded the archives in 1982, and the Library of Congress acquired it in 2002. A subset of the archives is freely downloadable and reusable at the Internet Archive.

Rick and Megan met in 1998 over a mutual interest in landscaping history (Megan was looking into restarting *Landscaping* magazine). They soon discovered that they shared, in Megan's words, "a remarkably similar set of cultural reference points and values, such as ... punk rock and the

Photography by Robyn Twomey

DIY movement. We shared similar ideas about collection-building, with regards to books. We both valued book collections very highly, but valued them as building blocks, tools, and raw materials, rather than as static objects to be locked away."

Soon, they — and their book collections — were married.

When the couple's collection, in Megan's words, "outstepped our immediate research needs," they started to think about public access. But the size of their bounty wasn't their main motivation for starting the Prelinger Library. That evolved out of a profound dissatisfaction with public and university libraries.

Public libraries have recently been shedding books to make room for elegant lounging spaces and rows of computers. And with their emphasis on query-based online cataloging, they are discouraging the sorts of serendipitous discoveries that occur while browsing the shelves. Meanwhile, academic libraries keep a lot of their most valuable research materials in "closed stacks," leaving the browser unaware of what she is missing.

As Megan told an interviewer for *In These Times*, "One of the ... barriers put in place by major research libraries is that they don't enable ordinary people to make use of extraordinary materials."

By contrast, the Prelinger Library provides extraordinary materials and then encourages serendipitous discovery. Megan has organized the flow of the materials into a sort of narrative structure.

Like a thoughtful DJ describing a playlist, Megan describes the interior taxonomic logic to her placement of books in an article on the library's website: "Row Two starts with what people do with what we pull out of the Earth: histories of manufacturing and industry. *Mill and Factory* rubs shoulders here with *Iron Age* and *Factory Management*. The next bank proceeds to how we move around the objects we've made with what we've pulled out of the Earth: histories of transportation infrastructure. Highways, cars, railroads, airlines, and even *Bus Transportation* magazine have their spaces here."

The extraordinary is the rule here (where else will you find the 1956–57 collection of *Modern Packaging* magazine?). The Prelingers don't believe that history is best served by reading contemporary books by historians, so the library is a treasure trove of original source material. Do you want to know how the U.S. government described aspects of its slavery policies in sordid detail? Then why not go directly to the source and read a government research publication

> "We value book collections very highly, but value them as building blocks, tools, and raw materials, rather than as static objects to be locked away."

on slavery from the mid-19th century?

Another unique aspect of the Prelinger library is its appropriation-friendliness. As with Rick's earlier film archives, the Prelingers want you to be able to take these fragments of history and use them for your own textual mashup — your own satire or commentary or research project. So they'll help and encourage you to grab text and visuals from, say, *Factory Management* magazine, *Soviet Woman*, *FBI Law Enforcement Bulletin*, or the complete 125-year run of the *Official Gazette of the U.S. Patent and Trademark Office*, and then scan and use them.

I point out that nobody gets in my face at my local library when I use their copier to copy pages out of their books. So what's so appropriation-friendly about the Prelingers' place? Megan responds: "While it may be possible to still make copies at the public library, that doesn't address the climate of fear that generally governs people's relationships to the use of published works. People have been conditioned by the copyright notices at Kinko's and other places to believe that they are not entitled to make something out of a published work. We encounter that with visitors who say, 'Is it really OK to make a scan of this? Really?'

"By offering our space as a workshop and encouraging people to photograph, copy, or scan, we create an environment where the use made of the materials is foregrounded in a way it isn't at the public library.

"Also, our library has a much greater proportion of public-domain materials than is present in a general reading collection in a public library. And it has been cultivated to be image-rich, and therefore of particular interest to artists. Combine that fact with the presence of a flatbed scanner, and the result is quite different than the fair-use xeroxing

BROWSER'S PARADISE: The Prelinger Library's vast offerings are wide open to the public, the rows of shelves inviting serendipitous discovery. Patrons are encouraged to use the space as a workshop, with access to a flatbed scanner.

of contemporary, copyrighted works at the public library."

The Prelingers facilitate open access to printed text in other ways as well. They've partnered with the Internet Archive to digitize 4,000 items in their library. And they actually take the trouble to investigate the public access possibilities of individual books. Says Rick, "The majority of works published from 1923 to 1963 are public domain, but the only way you can know is to check the physical copy for the proper form of copyright notice and look for a copyright renewal in the records of the U.S. Copyright Office. This process is hard to automate, but for books we use an excellent database built by Mike Lesk and optimized by the Stanford University library. For other items we consult our own copies of Copyright Office records. It is, for sure, a slow and artisanal process."

In defense of the public domain, the Prelinger Library has joined the Internet Archives in challenging recent changes to copyright law — changes that automatically copyright for 50 years all text made public. This new copyright law creates an "orphan class" of creative works, because even if authors no longer care about protecting their commercial rights, the copyright is automatically in force and permission for reuse can only be granted by the author, who may be dead or impossible to locate.

Before leaving, I ask the Prelingers if their library is replicable. "Yes!" says Megan. "We really encourage people to create their own idiosyncratic public-access libraries."

Imagine if all the most rabid collectors of books and odd (or not-so-odd) textual ephemera started merging collections in locales all over the world, and artfully organized them so that each one represented a way of seeing the world. Sounds pretty cool, right? Let a million public-access libraries bloom!

» Prelinger Library online collection:
archive.org/details/prelinger_library

» Prelinger Archives online collection:
archive.org/details/prelinger

R.U. Sirius is a notorious technoculture iconoclast. He recently launched the Open Source Party and QuestionAuthority — two incipient political organizations at mondoglobo.net.

Meet TRIZ

A methodology for inventive problem solving. By Richard Langevin

TRIZ founder Genrikh Altshuller.

I discovered the Theory of Inventive Problem Solving, or TRIZ (its acronym in the original Russian), when I began working with Lev Shulyak in 1992. He had translated an introductory text on TRIZ from Russian into his broken English, and he wanted help editing and marketing it.

In that manuscript I saw a fascinating concept that had global impact: a methodology that can help everyone from children to rocket scientists solve problems more quickly and with better results.

I signed on, and the decision changed my life. Soon after, Shulyak and I formed the Technical Innovation Center, dedicated to TRIZ. And the rest is history.

TRIZ (pronounced "trees") is the brainchild of Russian scientist and engineer Genrikh Altshuller. Born in 1926, he made his first invention at the age of 14 and was later educated as a mechanical engineer.

In 1946, he developed his first mature invention, a method for escaping from an immobilized submarine without diving gear. This invention was immediately classified as a military secret, and Altshuller started working in the patent office of the Caspian Sea Navy, where he became intrigued by the meta-question of how innovation happens.

Altshuller searched scientific libraries, but could not find even the most elementary book on the subject of inventing. Scientists have often claimed that their inventions resulted from accidents and serendipity, but Altshuller didn't accept this mantra. He decided that if a *methodology* for inventing did not exist, one should be developed.

Taking an empirical approach, Altshuller began by reading patents, a resource that was readily available where he worked. Patents represent the best documentation we have for what constitutes invention; they're detailed descriptions of new solutions to existing problems.

Grounded in the patent base, but abstracting beyond the specific technical subject matter, Altshuller found that a meaningful invention is, in essence, nothing more than the *removal of a contradiction without compromise*.

What does this mean? Basically, it means eliminating a technical or physical conflict in one characteristic of a system without compromising other characteristics of that system.

Further, Altshuller observed that the same problem types appeared time and time again, and yielded to correspondingly generic solutions. Innovation seemed to follow a taxonomy, which supported the idea that it could proceed methodologically, like other engineering disciplines. And once its techniques had been discovered and codified, they could be taught to engineers and children alike, and applied to generate innovative ideas for diverse problem situations.

Altshuller assembled a team and proceeded to develop this methodology. TRIZ knowledge is divided into two groups: Key Concepts and Tools. Key Concepts include Levels of Innovation, Ideality, Contradictions, and Evolution of Systems. Tools include the 40 Principles, Standards, Substance-Field Analysis, and ARIZ, the algorithm for solving nonstandard problems. As I tell my trainees, TRIZ is the natural way that creative people have solved inventive problems.

I believe TRIZ can help many makers become more effective problem solvers, and I hope you'll want to learn more about it. For more information, please visit our website at triz.org, or the Altshuller Institute for TRIZ Studies, a not-for-profit group dedicated to promoting and overseeing TRIZ development, at aitriz.org.

Richard Langevin is CEO of the Technical Innovation Center, Inc. (triz.org).

Maker

Fun Idea Machine

Proof that you can buy happiness ... for 24 cents.

By Jake Bronstein

When people see my Fun Idea Machine (a vending machine I've filled with plastic bubbles, each containing an idea for having fun) they usually ask me the following questions:

1. Where did the idea for the Fun Idea Machine come from?
I consult for ad agencies developing fun ideas to market products. So the idea of a vending machine that could sit behind my desk and "sell ideas" seemed like the kind of thing executives visiting my office would enjoy.

2. How did you build the machine?
Toy dispensers are pretty cheap: $80 brand new on eBay. From there, all I had to do was make the signs to stick to the inside glass. Originally it read, "Fun ideas: $1,000 and up. Consumer engagement: Priceless!"

3. So what's in the machine?
At first, nothing — it was just a prop. Clients would come in and ask if they could use it, and I would chuckle, "Sure ... it runs on quarters ... you're going to need 4,000." Everything changed when I brought it out to the street to take some pictures for my business cards. People kept stopping me, asking if they could use it.

4. So there's nothing in it?
Au contraire. I realized I was onto something. I rushed back to my office, ordered a box of toys and the eggs to put them in, and went to work filling it up. Obviously I had to change the price — the machine requires two quarters — but even that seemed excessive. So in addition to putting a fun idea in with each toy, I put in a quarter, and a penny

that you could leave heads-up for someone else to find (nothing is more fun than making someone's day). Put in 50 cents, and you get back a toy, a fun idea, and 26 cents. Fun, right?

5. What are your favorite ideas?
Having taken the machine out now about 30 times, made more than 400 ideas to put inside, and sold almost 4,000 "Funballs," I can tell you my favorites come in two categories. The first are things people can do right now, things that if they try, their friends will see and say, "I wanna do that." Like the "shaky face" or "high jump" (search for these terms at zoomdoggle.com to see what they are).

The second are things they might not do, but they remember a time when they would have, a time when they knew that fun was just something you do, regardless of the time or place. Somehow adults forget — they "make fun of," which is taking the fun out of, instead of "having fun with."

6. Are you going to start a Fun Idea Machine business?
I got a call from a vending machine business owner. He wanted to know how much it would cost to franchise the idea. I told him it'd be cool to have 10 machines out there and make a couple bucks ... but it'd be cooler to have 1,000 machines even if it meant people were stealing the idea. I don't have the patent on fun. The machine is just a fun thing for me *and* the people who use it. The funnest idea of all is to make one of your own. I promise you won't regret it.

Jake Bronstein was one of the founding editors of *FHM* magazine's U.S. edition. He keeps a blog about fun called zoomdoggle.com and can be found at jakebronstein.com.

Make: SPY TECH

A ny secret agent worth a bowtie camera knows that spying and technology are as intertwined as M and Q. This issue snoops into the 21st century with an invisible ink printer, a cellphone spy scope, a talking booby trap, a USB drive hidden inside a battery, and other stealthy schemes. The only question that remains is which project to start first. Now *that* takes intelligence. »

TALKING BOOBY TRAP

SURPRISE ENEMY SPIES
WITH THE SOUND OF YOUR
OWN VOICE.

M016

BY BOB KNETZGER

THE TELL-TALE TRAP: You set the trap, they get the scare. "Villain! Admit the deed!"

Having trouble with people snatching your top-secret stuff? Need help getting some privacy? Here's a sneaky gizmo you can make to keep those snoops away. It's a Talking Booby Trap: record your personalized message or sound effect, then hide it in a strategic place. When it's disturbed, the intruders will hear your surprise warning message telling them to "Get lost!"

Photograph by Garry McLeod

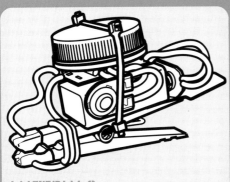

MATERIALS

9V Recording Module, **RadioShack part #276-1323.** This is a pre-wired record/playback unit complete with audio board, speaker, and controls that will record up to 20 seconds of sound in nonvolatile memory. There are actually 2 different versions of this part; try to find the one with the separate wired microphone. It's much better, with a louder and clearer sound!

9-volt battery

Clothespin serves as a spring-loaded booby trap and as the switch that triggers the sound

Rubber band

Aluminum duct tape

Wire

Plastic cap from 1gal milk jug

Super glue aka cyanoacrylate glue

Zip ties

Double-sided foam tape

Soldering iron, solder

Wire cutters

Small bit of brass strip (optional) with tiny screws (2)

Drill (optional)

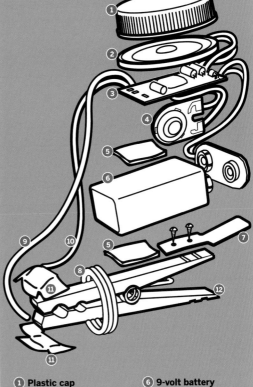

1. Plastic cap
2. Speaker (from module)
3. Module circuit board
4. Record button
5. Double-sided foam tape
6. 9-volt battery
7. Brass strip
8. Rubber band
9. Play wire 1
10. Play wire 2
11. Aluminum duct tape
12. Clothespin

WHOOP! WHOOP!

YOU'RE BUSTED!

Illustrations by Bob Knetzger

BUILDING THE BOOBY TRAP

1. Disassemble the clothespin — the wire spring is too stiff for our purpose. Reassemble the pin as shown in Figure A (next page) so that the spring is used as a fulcrum pivot. Wrap a rubber band around the jaws of the clip a couple of times. Slide the

rubber band closer or farther from the fulcrum to adjust the tension. The clip should open easily, but still be able to close all the way.

For an easier-to-use booby trap, add an optional extension to the top leg of the clothespin. Bend a small strip of stiff brass so that it lies flat when

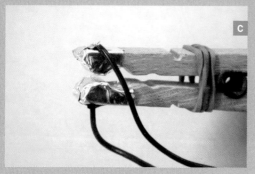

TO PINCH A THIEF: Fig. A: Take apart and reassemble a clothespin. Fig. B: Add an optional brass extension to the clothespin.

Fig C: Metal tape on the ends of the clothespin serves as contacts for the spring-loaded switch. Fig. D: A small plastic cap improves the speaker's sound.

the clothespin is held open. Drill 2 small holes in the brass and attach it to the clothespin with tiny screws (Figure B).

2. Wrap the jaws of the clothespin with the aluminum tape. Poke a small hole in the aluminum and attach the stripped end of a short wire (about 5" long) to each jaw. Twist the wire and crimp the tape over it firmly to make a good electrical connection (Figure C). You'll be soldering the other ends of these 2 wires to traces on the recording module's circuit board.

3. Install the 9-volt battery and test the circuit: press and hold the Record button (the red LED goes on) and speak loudly into the microphone. Release the Record button when finished. Press the gray button on the PC board to hear your recording.

You can improve the sound significantly and protect the naked speaker by adding a resonant chamber/cover. Use a plastic bottle cap from a gallon milk jug — it's just the right diameter. Super-glue it to the front of the speaker (Figure D).

4. Now modify the circuit to wire up the clothespin. Remove the battery. Find the gray rubber-domed Play switch on the PC board (Figure E). Bend the 3 metal tabs on the back and remove the button.

5. Feed a wire from the clothespin jaw through 1 of the tab holes and over the edge of 1 of the 2 switch traces. Carefully solder the wire to the first trace only — don't short out the traces! Do the same thing for the second jaw wire and solder it to the second trace (Figure F).

Now the clothespin will act as a Play switch: when the jaws touch together, the sound plays. Try it! To prevent the sound from playing, place a slip of paper between the jaws as an insulator.

6. To finish, stick the battery to the top of the clothespin with double-stick foam tape, and then stick it to the circuit board with more foam tape. Foam-tape the speaker on top, then use cable ties through the center of the spring to cinch everything together. Tuck in the microphone and Record switch wires to neaten it all up (Figures G and H).

Photography by Bob Knetzger

Fig. E: The sound recording module we'll be using. Fig. F: Remove the Play switch button on the circuit board. Feed the clothespin wires through 2 of the tab holes. Fig. G: Booby trap in set position. Fig. H: Booby trap in triggered position. Adjust rubber band tension as needed.

SETTING UP THE BOOBY TRAP

After you record your own message and sound effects, you can use the Talking Booby Trap in lots of ways:

» Place a diary, journal, or any object on the brass tab. The weight of the object keeps the clothespin open and armed. Camouflage the booby trap by placing something else in front. If anyone lifts the book — whoop! whoop! — the alarm goes off. The snoop is busted!

» Use the Talking Booby Trap to shame your lunch-lifting co-workers. Hide it behind your food inside the fridge at work. If anybody touches it, the talking booby trap blasts: "Keep your hands off my lunch!" Now everybody will know the identity of the secret food thief!

» Tie a string to a small piece of paper and slip it between the clothespin jaws. Then tie the other end to any pilferable object. You'll know if anyone tries to take it, because pulling the string trips the booby trap: "Step away from the camera!" You can use thin monofilament fishing line instead of string for a nearly invisible alarm.

» Any bit of paper can act as the bait in your trap to catch a thief.

» Open the Talking Booby Trap and slip it under a closed door. If anybody opens the door, you'll know it. "Hey, you kids — get back in bed!" Or record a scolding "Bad dog!" message to keep pets in their place while you're away. You can also arm a drawer or sliding door.

» On a more positive note, you can leave a friendly audio reminder for someone special. Arm their cell-phone or car keys and they'll hear you say: "Don't forget — romantic dinner tonight!" They'll really get the message from the spy who loves them!

Bob Knetzger (neotoybob@comcast.net) is an inventor/designer with 30 years' experience making fun stuff. He's created educational software, video and board games, and all kinds of toys, from high-tech electronics down to "free inside!" cereal box premiums.

PORTABLE SPY SCOPE

TURN YOUR CELLPHONE
INTO A HIGH-POWER DIGISCOPE.

M016

BY ERIC ROSENTHAL

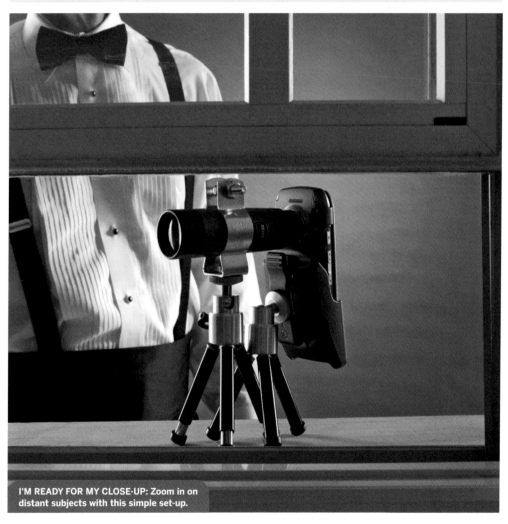

I'M READY FOR MY CLOSE-UP: Zoom in on distant subjects with this simple set-up.

Digiscoping is the process of coupling a digital camera to a spotting scope to photograph distant subjects. Digiscopes are popular with hunters and birdwatchers. By shooting digital photos through a powerful scope, you can produce highly magnified photographs with the same image quality as that of a camera with a 3,000mm–4,000mm focal length lens.

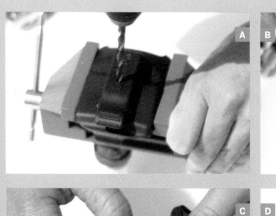
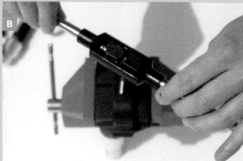

Fig. A: Drill a ¹³⁄₆₄" pilot hole on the cellphone holster's spring clip. Fig. B: Use a ¼-20 tap for the tripod camera mount screw.

Fig. C: Attach the mini-tripod to the cellphone holster with the mount screw. Fig. D: Mount the monoscope using a 1" pipe clamp.

MAKING A SPY SCOPE

I set up my system using 2 ball-head mini tripods. One tripod holds the cellphone (camera) in position and the other holds the monoscope. Depending on what kind of cellphone and monoscope you use, you may have to modify the setup described here.

I found a cellphone belt clip that fit my phone without covering the camera lens. I drilled a ¹³⁄₆₄" pilot hole on the spring clip (Figure A) and then used a ¼-20 tap so that the tripod camera mount screw would thread into the tapped hole in the belt clip (Figures B and C).

MATERIALS

Cellphone with a camera
Ball-head mini tripods (2) **I used Vanguard VS-55 mini table tripods but there are many other options.**
Cellphone belt clip **that fits your phone without covering the camera lens**
¹³⁄₆₄" drill bit and ¼-20 tap
Monocular telescope **aka monoscope**
1" pipe clamp **to mount the monoscope**
¼-20 nut

I mounted the monoscope using a 1" pipe clamp (Figure D) and attached the clamp to the ¼-20 tripod screw with a ¼-20 nut (Figure E, next page).

Once you've successfully mounted your cellphone and the monoscope on the mini tripods, find a stable, vibration-free platform large enough to hold both tripods.

Set the monoscope with its tripod on the platform. Aim and focus the monoscope on a distant object. Then position the cellphone with its tripod so that the lens of the cellphone camera is directly behind and in line with the eyepiece lens of the monoscope.

Turn the cellphone camera on and carefully position it so that you get the best image that fills the screen. Take the time to find the correct position; this is critical to obtaining quality images.

If you can't position the camera lens close enough to the monoscope eyepiece to fill the frame, then remove the rubber eye guard on the monoscope (it should just pull off) and try again.

You're now ready to take the photo. It's important to reduce vibration to a minimum when capturing the image. If your cellphone has a self-timer, set the timer to capture the image. If there's no self-timer

Photography by Ed Troxell

Fig. E: Attach the pipe clamp to the mini-tripod using a ¼-20 nut. Fig F: Tighten the nut on the pipe clamp to secure the monoscope.

Figs. G and H: A mailbox that's almost invisible to the naked eye is no problem for the spy scope. Figs. I and J: Chemical refinery or theme park? The spy scope knows!

then you have to be careful to keep the cellphone very steady when taking the picture.

Once you've determined the correct positioning of your camera and monoscope, you can construct a customized mount to make the rig more portable without the need for constant alignment.

Above are some images taken with my spy camera (Figures H and J).

More information and many more photos can be found at creative-technology.net/SpyCamera.html.

Eric Rosenthal is president of Creative Technology, LLC (CTech), a company specializing in new and advanced imaging technology, consulting, and development.

Photography by Ed Troxell (E and F) and Eric Rosenthal

SIMPLE LASER COMMUNICATOR

TALK IN SECRET
OVER YOUR PRIVATE
LASER BEAM LIGHT LINK.

M016

BY SIMON QUELLEN FIELD

LIGHT CONVERSATION: No one will suspect you're communicating through a laser.

ow would you like to talk over a laser beam? In about 15 minutes you can set up your own laser communication system using a cheap laser pointer and a few parts from RadioShack. The audio signal from a microphone varies the power feeding the laser, so that its brightness changes, following the shape of the original sound wave. At the receiving end, a solar cell or photo-resistor converts the oscillating light signal back into the original sound.

The communication is completely private, with no wire connection to tap into. Only you will be able hear what comes over the secret laser link.

Photography by Garry McLeod

Make: **67**

Make: SPY TECH

MATERIALS

Laser pointer You can get one for $10 from my Scitoys Catalog, scitoyscatalog.com.

Batteries, 1.5V, any size Get the same number of batteries and voltage that the laser takes, typically 3x 1.5V.

Battery holder for the 1.5V batteries If you can't find a 3-battery holder, wire a 1-cell and a 2-cell holder in series, or use a 4-cell holder and bridge 1 compartment with wire.

Stiff wire or rubber band

Audio output transformer, 1kΩ primary coil, 8Ω secondary RadioShack part #273-1380 or Scitoys Catalog #XFORMR

Alligator clip leads (2–10) with points fine enough to connect to the inside of the laser pointer. You can substitute wire and solder, but the clip leads are easier. RadioShack #278-1156 will do nicely.

2-lead bicolor LED to protect the laser from voltage spikes if it doesn't have built-in protection.

Hookup wire

Mini portable amplifier such as RadioShack #277-1008. Alternately, you can use a stereo system.

Microphone with cable and plug that fits amplifier or stereo input

RECEIVER OPTION #1: SOLAR CELL WITH EARPHONE (SIMPLEST)

Small solar cell Scitoys #3SOLARCELLS or RadioShack #276-124

Piezoelectric earphone Scitoys #EARPHONE

Transparent tape

RECEIVER OPTION #2: PHOTOCELL WITH EARPHONE (CHEAPER AND STURDIER)

Piezoelectric earphone Scitoys #EARPHONE

9V battery and battery clip

Photoresistor, cadmium sulfide (CdS) RadioShack #276-1657

220Ω resistor

TOOLS

Soldering equipment

Transistor radio

Earphone plug to fit your radio, such as RadioShack #42-2434

Phono/mic plug to fit the phono/mic input jack of your mini amplifier (or stereo), such as RadioShack #42-2434 or #42-2457

ASSEMBLE THE TRANSMITTER

NOTE: I recommend soldering this project, but initially it's easier to make it and test it using alligator clip leads.

Remove the batteries from the laser. Connect the external battery pack to the laser's power contacts with 2 alligator clips; usually you'll connect one lead to the battery case and the other to the spring inside. Some laser pointers are easy to disassemble; you can remove the circuit board and see the power

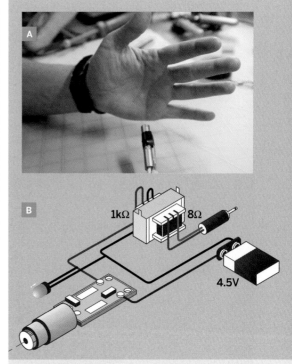

Fig. A: The laser pointer with its button taped down to the always-on position.

Fig. B: The transmitter wiring diagram.

contacts conveniently marked with a plus and a minus. If it doesn't light, try reversing the power; this won't harm the laser.

Figure out how to hold the laser's button down with a rubber band, wire, or tape (Figure A).

Remove the batteries. Following the schematic in Figure B, connect the 1,000-ohm (1kΩ) side of the transformer in-line between the battery pack and the laser, using the outside 2 wires of the transformer. We don't need the center tap wire.

Connect the bicolor LED between the same 2 wires of the transformer. This protects the laser from high voltage spikes, since so many cheap lasers nowadays have no onboard protection circuit. If you see the LED flash, that indicates a spike.

Now connect the earphone plug to the 8Ω side of the transformer with alligator clips. That's it! We have a laser transmitter, in just a few minutes! We'll plug it into a transistor radio for testing (Figure C).

ASSEMBLE A RECEIVER

For the simplest receiver, connect the 2 leads from a piezoelectric earphone to a small solar cell. You can attach them using transparent tape instead of

Illustration by Julian Honoré/p4rse.com

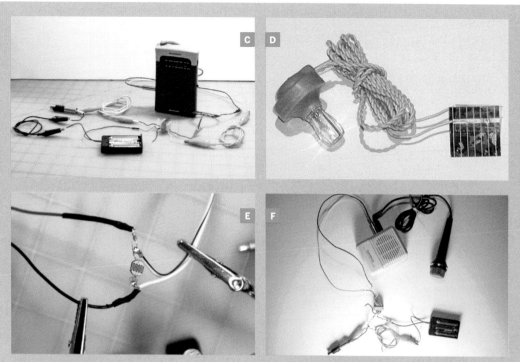

Fig. C: Transmitter ready for testing with a radio.
Fig. D: Simplest receiver: just an earphone and a solar cell.

Fig. E: The photocell (i.e. photoresistor) uses battery power. Fig. F: The working transmitter is fed by a microphone.

soldering, which can be difficult on solar cells (Figure D).

For a sturdier and cheaper alternative, use a cadmium sulfide (CdS) photoresistor, which changes its resistance proportional to the amount of light hitting it. Paired with a battery, this acts like a solar cell. To make the receiver, connect the earphone and a 9V battery across the photoresistor, so that battery, earphone, and photoresistor are all parallel. Add a 220Ω resistor in series with the battery to reduce power consumption and prevent heating of the photoresistor (Figure E).

SETUP AND TESTING

We'll test the system by transmitting a radio signal and amplifying the receiver so we can hear it across the room. First, replace the earphone of your receiver with an audio plug (or just clip the audio plug to the earphone wires), and plug it into an amplifier or stereo.

With the radio off, plug in the transmitter. Turn up the volume on the amplifier until you hear a hiss, then turn it down until it isn't noticeable.

Aim the laser across the room so it hits the solar

cell or photoresistor. You may hear some clicks or pops. Now turn on the radio and adjust its volume until you hear it across the room. If you don't hear it, try increasing the amp's volume before you turn up the radio. If you pull out the earphone plug, the radio should be just audible.

Depending on your signal source, you also might want to reverse the transformer. Some devices, like iPods, don't have enough power to drive 8Ω speakers, so you should connect them across the 1kΩ side. This arrangement will dim the laser, but won't affect its range much.

When you can hear the radio, break the laser beam with your hand, and notice that the music stops. Try chopping up the audio with your fingers.

The system is ready. To send secret voice communications, move the amp from the receiver to the transmitter and plug in a mic. You're ready for the field; just be careful with the volume, to protect the laser.

Simon Quellen Field (sfield@scitoys.com) is the president and CEO of Kinetic MicroScience (scitoys.com), and the author of several books on science and computing.

Photography by Ed Troxell (A, C and F) and Simon Quellen Field (D and E)

SURVIVAL SYSTEMS: SHAKEN, NOT STIRRED

MINIATURIZE A
SERIOUS SURVIVAL KIT.

M016

BY THOMAS AREY

CURIOUSLY STRONG SURVIVAL: It won't help your breath but it could save your life.

Q was not happy about the latest memo from M: budget cuts indicated that Q Division had to tighten its belt. No more custom-built Aston Martins for the men and women of the 00 Division. Q was now expected to outfit the agents with "off-the-shelf" devices to help them out of difficult situations.

He had great plans for his latest system, a survival kit that had all the essentials, including a shelter and a satellite transponder all built into an agent's shoe heels — but it looked like that project would have to be put aside. Agent 007 was to leave on a mission in the morning and he required a practical survival kit that would fit in the pocket of his evening jacket. Q took a mint tin from Miss Moneypenny's desk and set out putting together the essential tools for staying alive until help comes.

You don't have to be Q to figure out how to make

a small but truly useful personal survival system. With planning, careful shopping, and a bit of good old maker scrounging, you can pull together the essentials to keep you going until help arrives.

As someone who likes to get out into the woods, often solo, I understand the importance of choosing good survival tools. My favorite resource in such matters is Cody Lundin's book, *98.6 Degrees: The Art of Keeping Your Ass Alive*. Lundin shows how to assemble a custom survival kit that's more practical and better outfitted than those you can buy in a camping

Photograph by Garry McLeod

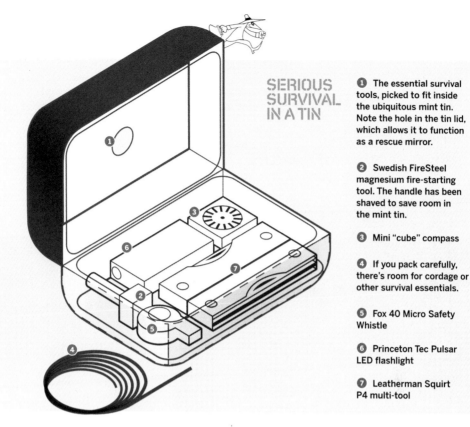

SERIOUS
SURVIVAL
IN A TIN

1 The essential survival tools, picked to fit inside the ubiquitous mint tin. Note the hole in the tin lid, which allows it to function as a rescue mirror.

2 Swedish FireSteel magnesium fire-starting tool. The handle has been shaved to save room in the mint tin.

3 Mini "cube" compass

4 If you pack carefully, there's room for cordage or other survival essentials.

5 Fox 40 Micro Safety Whistle

6 Princeton Tec Pulsar LED flashlight

7 Leatherman Squirt P4 multi-tool

or outdoor store. Stir in a little maker attitude with Lundin's experience and it's easy to build a pocket-sized survival kit that rival Q's best efforts.

I started the project with a mint tin. Since it's not waterproof, I needed to choose tools that could still do their duty if exposed to moisture.

Any good survival system begins with a high-quality cutting edge. I chose the Gerber Mini Paraframe knife ($12). It's made of stainless steel, has a 2¼" serrated blade, and weighs only 1½oz. You may choose to substitute a MAKE: Warranty Voider, aka the Leatherman Squirt P4 multi-tool (2¼", 2oz, $40 from makershed.com).

In the event I find myself in the dark, I included a Princeton Tec Pulsar LED flashlight (1", ¼oz, $8–$11). Its lithium batteries last for 12 hours.

Letting rescuers know where you're located can be as easy as whistling. In this case I've added a flat, "pealess" Fox 40 Micro Safety Whistle (2", ½oz, $6). And keeping your bearings can be critical, so I added a mini "cube" compass (1", ⅕oz, $2).

Being able to make a fire can help rescuers locate you, as well as provide necessary warmth. Since I wanted to keep this kit small and free from wet worries, I chose a Swedish FireSteel magnesium fire-starting tool (3", 1oz, $12). Lighting a fire is as simple as building a pile of dry tinder and striking the steel with your knife.

TIP: Save the sharp edge of your knife blade. The back edge will usually work just as well when striking a metal match.

If you carefully remove as much extra plastic or metal from survival devices as possible, you can stuff even more items into the mint tin. I was able to remove enough "meat" from the whistle, compass, LED light, and metal match to make room in the case for 12' of 7-strand nylon "paracord." This 500lb-test cord can be used whole, or the individual strands can be separated for other uses.

Not only can a mint tin serve to carry my survival essentials, its shiny inside surface makes for an excellent signal mirror. I just punched a small hole in the lid. Sighting through this hole allows me to flash the mirror onto my rescuer's location and draw attention to myself.

Your personal preferences and needs will dictate changes to this kit. As always, a bit of scrounging and repurposing will go a long way. And finally, the maker mindset will always be the most important tool you can bring to an emergency or survival situation. Q would be proud.

Thomas Arey has been a freelance writer to the radio/electronics hobby world for over 25 years and is the author of *Radio Monitoring: A How-To Guide.*

DEAD DROP DEVICE

A HOLLOW BOLT
HIDES SECRET MESSAGES.

M016

BY BRIAN DEREU

DEAD BOLT: Who would think this hardware is threaded with conversation?

When spies need to pass off a note or microfilm without meeting in person, they sometimes use a "dead drop" device. An effective dead drop device looks common enough to blend in without causing suspicion under the casual glances of passersby. This bolt is one such item.

It's made from an ordinary steel bolt that's hollowed out and fitted with a threaded, removable head. If made to the correct size and properly painted, it could replace a solid bolt on a fence, bridge, or similar structure. (Of course, we don't advocate that you remove a real bolt from a bridge! The device described here should be considered a desktop curiosity only.)

Photograph by Garry McLeod

Fig. A: Thread a nut on the bolt, put it in a vise, and cut off the head with a hacksaw. Fig. B: Clean up the bolt shank and head with a file or sandpaper.

Fig. C: Use a center drill on the shank of the bolt to prevent the twist drill from "walking." Fig. D: Hollow the shank of the bolt with a ⅜" twist drill.

MATERIALS

½-13×2" SAE Grade 1 or 2 steel bolt
Nut to fit bolt
Small O-ring or rubber seal (optional)

TOOLS

Vise
Hacksaw
File or belt sander
Drill press with vise
Center drill bit
⅜" twist drill
²⁹⁄₃₂" drill bit
Cutting oil
Countersink bit
½-20 thread, high-speed steel tap and die
Sandpaper
Grinder
Small, tapered center rod or steel rod
to grind your own

1. BEHEAD THE BOLT.

Begin by cutting off the head of the bolt with a hacksaw on a bench vise (Figure A), and then cleaning up the shank and the head with either a file or a belt sander (Figure B). The jaws of a vise will crush the threads of the bolt, so hold the bolt by a nut that is run all the way up the shank.

2. HOLLOW THE SHANK.

Transfer the bolt shank to a drill press vise and adjust it to bring the spindle of the chuck directly atop the center of the shank. When drilling steel, it's always advisable to start the hole with a center drill to keep the twist drill from "walking" (Figure C).

After center drilling, drill the cavity hole in the shank with a ⅜" twist drill to a depth of 1½" or so. Use plenty of cutting oil, and peck the drill up and down often to clean out the chips and to introduce fresh coolant into the hole (Figure D). Chamfer the hole with a countersink bit.

Photography by Brian Dereu

Fig. E: Thread the shank of the bolt down 3 threads. Fig. F: After threading, deburr and clean up the shank with a file and sandpaper.

Fig. G: Drilling and tapping the bolt head is tricky, because the head is shallow. Use the depth stop on the drill press. Fig. H: Make a bottom drill from a twist drill.

3. THREAD THE SHANK.

Now the shank goes back to the bench vise for threading. I chose a ½-20 thread for this project. When purchasing taps, dies, and twist drills, always choose high-speed steel over carbon steel, and buy the best you can afford.

Thread the shank of the bolt down 3 threads (Figure E). Some dies have a taper on 1 end, and some are tapered on both ends. If your die has a single-sided taper, use that end first, and then flip it around and clean up the threads with the non-tapered end. Deburr and clean up the shank with a file and sandpaper, and it's done (Figure F).

4. DRILL AND TAP THE HEAD.

Next, the head of the bolt requires drilling and tapping. Use a ²⁹⁄₃₂" drill bit for a ½-20 tap. In this step, it's imperative to set the depth stop on the drill press; set it to leave around ¹⁄₁₆" of metal at the bottom of the hole. As the bolt's head is already shallow to begin with, the twist drill will need to be modified into a bottom drill after the initial drilling (Figure G).

To grind a bottom drill, start by grinding the end of the drill flat and square. Continue by grinding the

back of each cutting lip approximately 20° from flat (Figures H and I). The idea is to produce a cutting edge and to keep only the lips in contact with the hole bottom.

After grinding the bottom drill and reinserting it into the chuck, start the drill press on its lowest speed, and don't start the spindle until the bottom drill is inside the hole. Use plenty of cutting oil, and feed slowly. When finished, the hole will be at full diameter throughout its entire length.

Without moving the bolt head or the drill press vise, replace the drill in the chuck with a small tapered center, which can be ground from a small metal rod (Figure J). Its purpose is to allow the tap to pivot on it, keeping it on center and square to the bolt head.

Using plenty of cutting oil, begin tapping the hole (Figure K). Hand taps come in 3 configurations: taper, plug, and bottom. A taper tap has several threads tapered for a very gradual cut, and is used mainly for through-holes. Plug taps have fewer threads tapered, and bottom taps have a taper on only the first couple of threads. For the few threads needed here, a bottom tap will suffice. As you advance the tap each

Photography by Brian Dereu (E, F, G, K) and Ed Troxell (H, I, J)

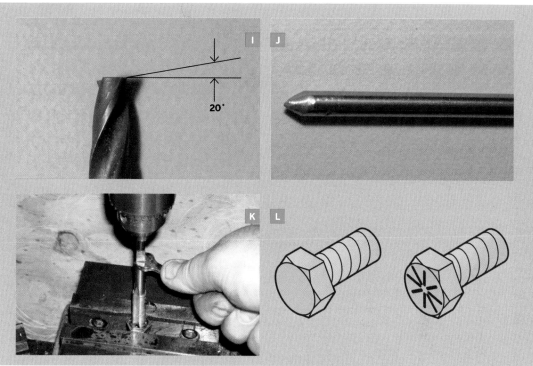

Fig. I: To grind a bottom drill, start by grinding the end of the drill flat and square. Continue by grinding the back of each cutting lip approximately 20° from flat.

Fig. J: Make a tapered center from a small metal rod. **Fig. K:** Tap the hole in the head. **Fig. L:** Hash marks on bolt heads indicate the grade of steel. Use Grade 1 or 2.

quarter turn, back it out to break the chips and allow easier tapping. Continue until the tap bottoms out, then remove the tap and clean the hole.

5. TEST AND TWEAK.

Test the thread engagement between the bolt head and shank to see if the shank will thread at least 2 threads. If not, it may be necessary to grind off the first thread of the tap, enabling it to cut down deeper at full thread diameter as you re-tap the bolt head. Deburr and clean everything, and your spy bolt is complete.

Besides being used as a dead drop device, this hollow bolt can be used to conceal anything small enough to fit in its cavity. If there's enough room in the head, a small O-ring can be used to make the bolt waterproof. Keep the device protected with a light coat of oil when it's not in use. The bolt was likely plated when new, but machining it exposes the uncoated steel that will eventually corrode when exposed to moisture.

Brian Dereu is a self-employed manufacturer who enjoys gadgets, fishing, and family.

A BIT ABOUT BOLTS

The bolt for this project can be of many sizes. For ease of machining, it should be SAE Grade 1 or 2 steel; both are low-carbon and not heat-treated.

The head of the bolt tells the grade by a series of radial hash marks. No hash marks at all indicates the lowest grade, and therefore the softest and easiest to machine. Select a bolt that has a portion of its shank unthreaded (a long length-to-diameter ratio) which will leave a section available for the new threads.

For this project, I chose a ½-13×2" bolt. In bolt designations, ½-13 means ½" diameter, and 13 threads per inch. The length of a bolt is measured by its shank, not including the head.

Illustration by Julian Honoré/p4rse.com

THIS OBJECT WILL SELF-DESTRUCT...

CREATE OBJECTS THAT
MELT INTO USELESSNESS
AT YOUR COMMAND.

M016

BY ANDREW LEWIS

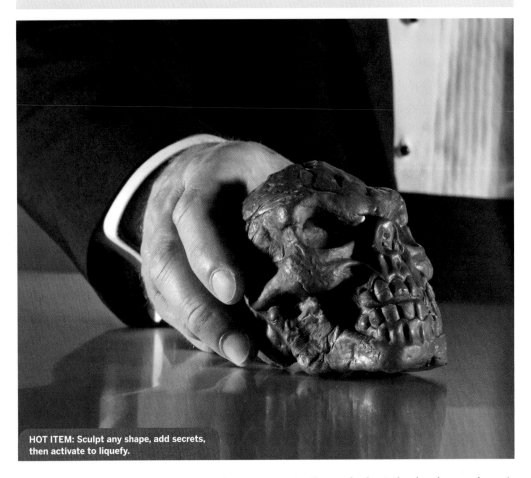

HOT ITEM: Sculpt any shape, add secrets, then activate to liquefy.

You've locked the top-secret documents in the safe, but the bad guys burst through the door before you have a chance to leave. What should you do? You could try swallowing the key, but these guys look like they might use a little amateur surgery to retrieve it.

The answer is simple — you calmly hand them the key and start edging your way toward the door. Seconds later, the head bad guy lets out a scream, and you flee the scene, leaving the intruders with a molten heap of plastic that used to be a key.

OK, so this scenario is a little farfetched, but it does neatly describe a use for this project. A battery, a reed switch, some Constantan wire, Friendly Plastic, and powdered aluminum are all you need to make a self-destructing object.

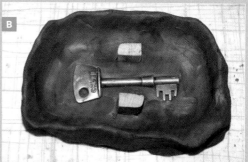

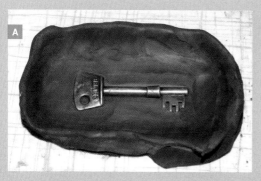

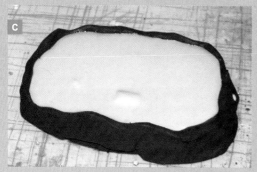

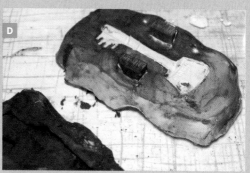

Fig. A: Choose an object to copy, like this key.

Figs. B–D: Make a mold using casting plaster (as shown here) or resin. Be sure to spray light oil or mold release into the mold before using it.

MATERIALS

CR2 lithium camera battery

Reed switch **normally open type**

Small piece of Constantan wire **A copper-nickel alloy, it's sold by electronics shops.**

ShapeLock or Friendly Plastic **aka Polymorph in the U.K. and Europe. All of these are polycapro-lactone (PCL) plastic, a biodegradable polymer with a low melting point.**

Powdered aluminum **sold by hobby or model suppliers**

Spatula or spoon **Wood is good; use something that won't melt and won't get hot in your hand.**

Brass curtain rings (2)

Electrical tape

Thin foam tape

TOOLS

Hot plate or alcohol burner

Heat gun or hair dryer

Gloves and goggles

Magnet

1. MAKE A MOLD OF THE OBJECT YOU WANT TO CAST.

Choose an object to copy (in this case a key) and make a mold from casting plaster or resin (Figures A–D). I don't intend to provide detailed information on how to make a mold, since Adam Savage did such an excellent job in MAKE, Volume 08 ("Primer: Moldmaking"). Just remember to spray some light oil or mold release into the mold before you use it.

2. MAKE AN ELEMENT WIRE.

The element wire should form a loop that passes through as much of the object as possible, so bend it into a suitable shape (Figure E, following page). The gauge and length of the wire will affect the temperature, as will the voltage and type of battery you use.

I chose a CR2 battery, and about 5" of 28-SWG Constantan wire. This battery and wire combination will produce a fairly dramatic result when activated, but it will also get hot enough to burn or even catch fire. For safety reasons, it's probably better to use either a lower-voltage battery or a thinner (cooler) gauge of wire.

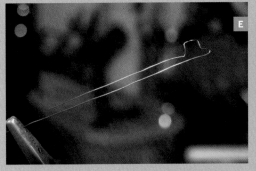

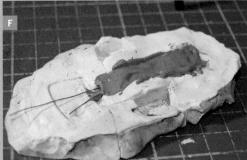

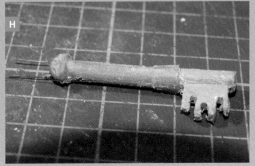

Fig. E: Bend the element wire into shape, forming a loop that passes through as much of the object as possible. Fig. F: Push the molten plastic into the mold, then lay the element wire over the top of 1 half of the mold. Fig. G: The finished molding will have excess bits attached. Fig. H: Clean off the excess with a sharp knife.

3. MIX PLASTIC COMPOSITE.

Friendly Plastic is a low-temperature, moldable plastic called polycaprolactone (PCL). At higher temperatures, PCL becomes sticky (like a hot glue), and it can be mixed with various additives to alter its physical properties.

To make the composite, simply melt some Friendly Plastic in a pot over low heat until it's easily stirred but not so hot that it smokes. Then switch off the flame and add the aluminum powder, at a ratio of about 2 parts plastic to 1 part aluminum. The aluminum will help transmit heat evenly through the plastic, accelerating the remelting process. Mix the composite until you get a smooth, gray emulsion.

! CAUTION: Be very careful with the plastic in its molten state — when overheated, it behaves a bit like boiling sugar, and it will stick to your skin, leaving a painful burn. Because you're melting the plastic over a direct heat source rather than with hot water, you should always handle it with a spatula and not with your hands.

4. CAST YOUR OBJECT.

Use a spatula to push the molten plastic into the 2 halves of the mold and then lay the element wire

over the top of 1 half of the object (Figure F). The ends of the wire should stick out from the end of the object, so that they can be connected to the rest of the electronic components.

Remelt the plastic in the mold with a heat gun, and then quickly bring the 2 halves of the mold together. The 2 plastic pieces should bond together, trapping the element wire between them. Let the mold cool for about a half hour, and then open it.

You may find that you need to smash the mold to get at the object. This usually happens if you stuffed too much plastic into the mold, or didn't use any mold release.

The finished molding will probably have excess plastic and bits of the mold clinging to it (Figure G). Clean these off with a sharp knife, and smooth any rough edges with a hot blade (Figure H). Be careful at this stage, since the plastic composite will melt very quickly if you apply too much heat.

5. ATTACH THE BATTERY AND SWITCH.

Next, you need to connect your battery to a reed switch. The easiest way to do this is to strip the plas-

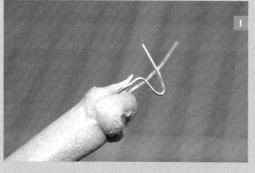

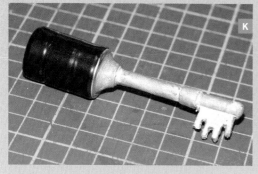

Fig. I: The ends of the wire should stick out. Fig. J: Poke 1 wire through the battery hole and twist the other onto the reed switch leg. Fig. K: Anchor the element and

battery and polish off the look with 2 small brass rings. Fig. L: Once activated, the key should begin to sag and melt within 10 seconds.

tic coating from the battery, and bend 1 wire on the reed switch back 180°. You can then tape the switch to the battery using ordinary electrical tape. Make sure you use enough tape to prevent a short circuit.

Most CR2 batteries have a small hole through the positive cap, and this can be useful for positioning the element wires. Poke one wire from the element through the hole, and twist it to secure. Twist the other wire onto the leg of the reed switch (Figure J). Be very careful not to short-circuit anything here, because a minor slip at this point can prematurely trigger the element and spoil all your hard work.

6. CONCEAL THE DESTRUCT CIRCUIT.

All that remains is to anchor the element and battery in position using more plastic composite, and to make the battery look less conspicuous. Simply glue 2 small brass curtain rings at either end of the battery, pad the sides with foam, and then wrap with the tape of your choice (Figure K).

7. SELF-DESTRUCT!

To make the key self-destruct, place it on a suitably

protected surface in close proximity to a magnet. The magnet activates the reed switch.

Once the self-destruct has been activated, the key will quickly heat up. The key should begin to sag and melt within 10 seconds (Figure L), and within 1 minute you'll have just a pile of molten plastic and a glowing red wire.

Remove the magnet as soon as the key has melted; if you leave it in place for more than a couple of minutes, the plastic will start to bubble and the battery will overheat.

! CAUTION: Molten plastic will stick to your skin and burn you, so wear appropriate safety equipment: gloves and goggles at a minimum. Remember to stand well back and remove the magnet as soon as the key has melted. Leaving the magnet in place for a prolonged period could cause the battery to overheat, and in extreme cases, explode.

Andrew Lewis is a keen artificer and computer scientist with interests in 3D scanning, computational theory, algorithmics, and electronics. He is a relentless tinkerer whose love of science and technology is second only to his love of all things steampunk.

USBATTERY

HIDE A SECRET
FLASH DRIVE IN AN
INNOCENT AA CELL.

M016

BY ANDREW LEWIS

**KNOWLEDGE IN POWER: This hollow AA cell conceals
a USB drive and uses a watch battery for "cover."**

K eeping secret documents out of enemy hands can be a challenge,
especially if the bad guys are in the habit of stopping and searching you.
Encryption is one thing, but it's the art of steganography (hiding messages)
that will ultimately save you from a small, hot room with no windows.

This project shows you how to make a USB flash
memory battery useful for storing secrets far from
prying eyes. The battery can store a gigabyte of
data, looks just like a normal AA alkaline cell, and
shows 1.5V if you test it. It'll even power small
electronic devices.

Photograph by Garry McLeod

MATERIALS

Discharged AA alkaline battery
Small USB flash drive **the smaller and thinner, the better. I used a Kingston DataTraveler 1GB.**
Inkjet-printable clear sticker film or sticker paper
USB Mini B socket **desoldered from an old camera or USB hub**
Thin, insulated wires **from an old network cable**
Small rare earth magnets (2)
Rapid-setting epoxy glue
AG13 button cell battery
USB A to USB Mini B cable

TOOLS

Craft knife
Junior hacksaw
Disposable gloves
Goggles
Countersink or drill with 12mm bit
Screwdriver or probe set
Metal file **if you need to trim the flash drive a bit**
Soldering iron and solder
Small clamps
Steel wool
Fine sandpaper
Computer and printer

1. REMOVE THE AA BATTERY'S PLASTIC SHEATH.

Cut off the plastic wrapper that covers the dead AA battery, taking care not to score the metal underneath. Depending on the brand, you might need to clean glue from the battery case using a label remover.

2. SCORE THE BATTERY AROUND THE BOTTOM END CAP.

Some batteries (Duracell, for example) have a small indent in the case near the negative terminal; make cuts here with your hacksaw. Don't cut through the battery with a single cut. Just pierce the outer case, then rotate the battery, making a series of small cuts (Figure A).

 CAUTION: Wear gloves and goggles when cutting and emptying the AA alkaline battery.

3. EMPTY THE BATTERY.

Remove the end cap of the battery, using pliers if necessary, and save it for later (Figure B). The inside is filled with chemical mush. Scoop this out and dispose of it safely. Wash the battery in cold water to get rid of any remaining mush.

4. REMOVE THE CARBON LINING.

Hold the battery in a cloth and use either a drill or a countersink to carefully and slowly grind out the carbon. A 12mm bit should fit inside the battery quite neatly (Figure C, following page). Stop every now and again to empty the carbon dust out of the battery casing (Figure D).

! **CAUTION: If you drill too fast, the casing will get too hot to hold, or the drill will clog with dust and snatch the casing from your hand.**

When the casing is empty, wash it again and clean it with a piece of cloth on a screwdriver. Carbon dust is electrically conductive, so you don't want to leave any in there to short out the USB circuitry.

5. CRACK OPEN YOUR FLASH DRIVE.

The plastic case on most USB thumb drives is clipped together and can be opened fairly easily by levering around the seams with a knife (Figure E).

The space inside the battery is tight, so you may need to file the edges of the USB circuit board and gently flex the metal casing to get it to fit. Just be careful not to knock any components off. Also look out for any "through-the-board" links, and make sure that you don't grind them away (Figure F).

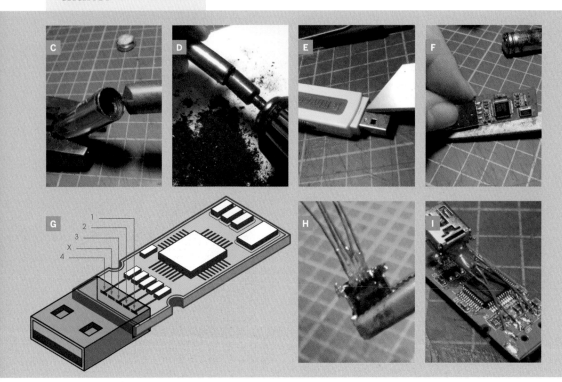

6. REMOVE THE USB PLUG FROM THE FLASH DRIVE.

The standard USB plug is too large to fit inside the battery, so it needs to be removed with a soldering iron or a micro heat gun. Heat the solder as evenly as possible, without melting any of the surface-mount components. Don't pull the socket; applying too much pressure will break the fine metal tracks on the circuit board.

7. WIRE THE TAGS ON THE MINI USB SOCKET.

The wires should be connected as shown in the diagram (Figure G). This will take a steady hand and plenty of patience. If you have a magnifying light, you might want to use it. Begin by applying solder to the wire and the socket individually, then put them together and touch them with the iron to briefly remelt the solder. Holding the iron on for too long will melt the socket. Once the wires are soldered in place, strengthen the joint by applying a little epoxy resin to the back of the socket (Figure H).

8. CONNECT AND TEST THE MINI SOCKET.

Trim the mini socket's wires so that they reach the tags that the original USB plug was connected to.

Solder them into place, but don't overheat them (Figure I).

> ⚠ **CAUTION: Make sure you've wired the socket correctly. It's possible to permanently damage your computer if you don't.**

Plug the USB drive into the computer to check that it's working. If it is, then carry on making; otherwise go back and check your wiring. The circuit must work before you can continue.

9. GLUE THE SOCKET AND WIRES IN PLACE.

Use more epoxy, holding the socket in place with a small clamp if necessary (Figure J). When the glue is dry, remove the clamp and slide the circuit into the battery housing, with the socket pointing out.

10. INSTALL 2 SMALL MAGNETS IN THE CASE.

Set these back slightly from the edge of the case, but don't block the USB socket (Figure K). Make sure the poles of the magnets are both facing the same way relative to the battery case.

Apply a little more epoxy and hold the magnets in place with small clamps or tweezers until the glue dries.

Illustration by Julian Honoré/p4rse.com

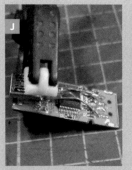

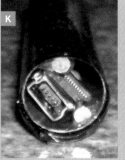

1.1. SALVAGE THE RING FROM THE END CAP OF THE AA.

You'll probably find a plastic plug with a thin metal cap stuck inside the end cap you removed from the AA battery. You need to remove this plastic. Pick it away with pliers, blades, or a drill (Figure L).

Once you've liberated the small metal ring from the plastic plug, clean the ring with steel wool.

1.2. INSTALL THE AG13 BUTTON BATTERY.

Melt some solder onto the inside of the ring from the AA, and then scuff the outside of the AG13 battery with fine sandpaper.

> ⚠ **CAUTION: Don't use steel wool to scuff the AG13 battery, as it will short-circuit the cell.**

Place the AG13 on a heat-resistant surface with its negative terminal facing up. Center the ring on the AG13 so that its negative terminal is positioned where the AA's negative terminal used to be. Melt the solder on the ring again to join them. If you can't get the solder to stick properly, use some spots of epoxy to reinforce the weld.

Now the end cap should fit back onto the end of the battery case, and be held in place by the magnets you glued in earlier (Figure M).

1.3. CREATE A NEW WRAPPER FOR THE BATTERY.

A suitable piece of artwork can be downloaded from makezine.com/16/usbattery and printed onto sticky-backed inkjet paper or film (Figure N). Before you stick it on, use a cotton swab and solvent to remove the adhesive from the bottom ½" of the label. This makes it easy to remove the bottom of the battery without it sticking to the artwork.

The battery should now look like an ordinary AA battery, and also register a normal 1.5V voltage if you check it with a voltmeter, thanks to the AG13 cell fitted to the end cap.

To remove the end of the battery and reveal the USB socket, simply attach a small magnet to the bottom of the battery and pull it away from the rest of the battery.

You now have your very own secret USB drive that you can use to smuggle those all-important Death Star plans past those pesky Imperial stormtroopers. To make it really blend in, you should cover some other normal batteries with artwork that matches the USB battery.

Andrew Lewis also wrote "Self-Destructing Object" and "Checkmate, Mr. Bond!" in this volume of MAKE.

CHECKMATE, MR. BOND!

UNLOCK A SECRET
COMPARTMENT WITH
MAGNETIC CHESS PIECES.

M016

BY ANDREW LEWIS

TREASURE CHESS: An ordinary-looking game board is the last place they'll look.

Everyone needs somewhere to stash stuff, and this chessboard provides a cunning hiding place whether you're a super-spy or a super-villain. Moving the chess pieces across the board in a particular way will unlock a secret compartment that you can use to hide diamonds, baseball cards, or the evidence linking you to the crime of the century.

Photograph by Garry McLeod

Fig. A: The chessmen after casting. Fig. B: Paint the chessmen. It seems more authentic not to varnish them. Fig. C: The completed set, behind a section of game board tiles. Fig. D: Two chessmen will contain magnets, which are used to open the secret drawer.

MATERIALS

10lbs or 5kg of high-strength molding plaster **such as Hydro-Stone or Denscal (Herculite Stone or Crystacast in the U.K.)**
Latex chess set molds **including chessmen and 2" square molds for the tiles**
Acrylic paint
Acrylic clear coat (optional)
Rare earth bar magnets (4) **small enough to fit inside the base of the chess pieces**
PVA adhesive **such as carpenter's glue or ordinary white glue**
¼" medium-density fiberboard (MDF) pieces:
　16" squares (2) **for the chessboard top and base**
　8"×12" **for the drawer base**
Dimensional lumber:
　1×2×16" (2) **for the chessboard sides. Remember that 1×2 dimensional lumber actually measures ¾"×1¾".**
　1×2×12½" **for the chessboard back**
　1×2×3⅛" (2) **for the chessboard front**
　2×2×14½" (2) **for the drawer guides**
Hardwood:
　1"×½"×8" (2) **for the drawer's front and back**
　1"×½"×12" (2) **for the drawer's sides**
　¼"×½" strip, 11½" lengths (2) **for drawer runners**
　2½"×¼" trim boards **I used reclaimed oak.**

Photography by Andrew Lewis

MAKE THE CHESSMEN AND TILES

1. Cast the chessmen and tiles.

Wash the latex molds in soapy water, and let any excess water drain out.

Mix the plaster with water and pour into the molds. The correct powder to water ratio for Herculite is about 2:1, but a good general rule is to aim for the consistency of thick pancake batter. Support the latex molds with jars or tins, so that the pieces don't fall over.

2. Remove the pieces from the molds.

Once the plaster is firm and dry (in an hour or so), you can remove the pieces from the molds (Figure A). You'll need to produce a complete set of chessmen from the molds, and also 64 tiles. Making the tiles might take a while, but it's worth the effort.

3. Paint the chessmen and tiles.

Once you've molded all the pieces, you can paint them with acrylic paint. The color scheme is up to you, but a more traditional look will probably attract less attention. Chess pieces tend to look best when

Fig. E: Chessboard with top removed, secret door closed. Fig. F: Secret door open. Note the drawer runners underneath the drawer. Fig. G: Close-up of the magnet slots.

Note the small wooden stop to keep the magnet from falling out. Fig. H: Remember not to glue the corner trim to the sides!

left unvarnished, while the tiles will benefit from a layer of clear acrylic (Figures B and C, previous page).

4. Install magnets in 2 chessmen.

You need to install bar magnets into 2 of the pieces, so that they can be used to activate the locking mechanism. The easiest way to do this is to hollow out the bottom of the chess pieces with a hobby drill or similar tool, and then glue the magnet in place before covering the hole with plaster (Figure D).

MAKE THE SECRET COMPARTMENT CHESSBOARD

1. Seal the MDF pieces.

Start by sealing the MDF pieces with a 50% solution of PVA adhesive and water.

2. Glue the wood sides to the chessboard base.

Glue boards all around the edge of the MDF base like the sides of a box, except for an 8¼" gap centered on the front (where the secret drawer will be). The thickness of these side boards isn't critical (I used ¾"-thick 1×2 stock), although a height of at least 1¾" is recommended. You can improve on

the strength of the glue with heavy-duty staples or tacks, but these aren't absolutely necessary.

3. Add the drawer guides and runners.

Cut a slot the same size as the bar magnets across the top of each drawer guide, then glue the guides into place on the base (Figure E).

A small plug of wood should be glued into the outside end of each slot to stop the magnets from falling out (Figure G). Also glue the ¼"×½" drawer runners to the base; these go under the drawer to make it easier to slide the drawer open and closed (Figure F).

4. Make a simple drawer.

Glue 1"×½" sides, the front, and the back to the 8"×12" MDF drawer base. I miter-cut the corners for strength and appearance.

To make the drawer look more presentable, you can either line it with material or apply a coat of paint. The drawer should slide freely between the drawer guides, and it should be the same height as the sides of the box (or nearly so).

Photography by Sam Murphy (E, F, G) and Andrew Lewis (H)

Photograph by Garry McLeod

5. Add magnetic locks to the drawer.

Carefully line up the drawer with the front edge of the box, and then glue some wooden stops behind to prevent it from slipping too far inside. Now mark the drawer sides where the magnet slots are, and cut a matching recess in the drawer sides. The bar magnet should be able to slide freely into and out of the slot in the drawer.

Place the bar magnets into the slots, and temporarily place the MDF top onto the box. You can now test that the magnetic locks will work by passing a magnet over the area of the locks, sliding the bar magnets back and forth in their slots.

6. Assemble the chessboard.

Remove the MDF top and glue the tiles to it using contact adhesive. Trim the edge of the box using hardwood strips. Make sure that on the front side you glue the trim to the drawer only, not to the box sides (Figure H). The trim should be wide enough to cover the box and the edge of the tiles.

Finish the corners with a wooden corner molding of your choice, but remember to glue it in such a way that the operation of the drawer is not affected.

7. Finish.

All that remains is to finish the wood trim and give everything a final polish. I used a dark wood stain and beeswax to finish the wood, and then gave it a polish with a soft cloth.

You could just as easily use metal or plastic trim instead of wood, but I prefer the more natural look that wood offers.

Andrew Lewis also wrote "Self-Destructing Object" and "USBattery" in this volume of MAKE.

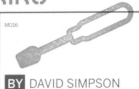

COVERT WIRELESS LISTENING

INSTALL A SNEAKY
BUG IN A BOOK.

M016

BY DAVID SIMPSON

THE SPY WHO BUGGED ME: Don't judge a book by its cover.

Connect a shirt-pocket "amplified listener" hearing aid with an in-car FM transmitter, and you've got a wireless bug.

Tuck them inside a hollowed-out book with the mic concealed by the dust cover, and you've got a covert listening device that you can leave lying around or on a shelf near a surveillance target. Then an undercover agent in the next room can eavesdrop on any devious plotting through her FM radio.

This is a fun and easy project, and if it weren't for the "covert" part, you could have all the components working together before you leave the RadioShack where you got them. (Really, I did!)

I used a book to camouflage my listening system, but you can also use a stuffed animal, a plastic toy (the big robotic WALL-E has potential), a basket of potpourri, or maybe even your dog's collar. The important considerations are:

» Make sure the microphone isn't obstructed.
» Provide an easy way to turn the power on and off.
» Don't enclose the electronics in metal, which limits the transmitter's range.
» Protect the delicate connections inside.

Photograph by Garry McLeod

IF THIS BOOK HAD EARS: Your sister might be listening to everything you're saying!

MATERIALS

Miniature "amplified listener" **I used RadioShack #33-1096; the Listen Up ("As Seen On TV") didn't work.**
Small wireless FM transmitter **I used the Accurian T707, RadioShack #12-2054.**
Personal music player with FM receiver **I used an old Walkman-style cassette player.**
Earbuds or headset
AA batteries (2)
2x AA battery holder
Hookup wire, red and black, around 22 gauge
Heavy shirt cardboard or illustration board
Heat-shrink tubing
Hardcover book
Velcro tape
White glue

TOOLS

Screwdrivers
Alligator jumper cables
Soldering/desoldering supplies
Hobby knife
Wire cutters/strippers
Ruler
Binder clips, large (1–2)
Drill and ½" drill bits
Paint brush, about ½" wide
Marker
Wood boards (2), with drywall screws (4) or handscrew clamps (2) **for use as a book press**

TESTING, ONE, TWO, THREE

Let's launch this mission. First, we'll test everything. Plug the headphones into the listener and turn it on. You should hear your surroundings like a bat.

Now plug the headphones into the FM receiver and tune it to a static-only spot at the low end of the dial, where these wireless FM transmitters generally broadcast.

Plug the transmitter into the listener, tune it to match the receiver, and you should be able to hear the sounds of the listener through the receiver. The signal should be strong, but it still might have some static.

Have a co-conspirator help with your test by talking near the listener while you walk away, listening through the receiver. See how far you can get before the signal is no longer intelligible. (For the best range, make sure to use fresh batteries.)

Test the components individually and together like this after each step of the assembly process. Delicate wires can snap and tiny switches can get switched at any stage, making troubleshooting difficult.

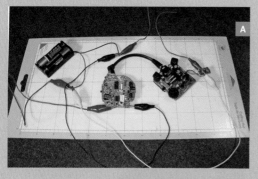

Fig. A: Components clipped together for initial testing.
Fig. B: Mics extended from listener PC board.

Fig. C: The listener and transmitter share connections to the battery pack. Fig. D: Cut through the pages to hollow out the book.

GUT AND EXTEND

Shed the extra bulk by unscrewing the listener and transmitter's plastic housings and taking out their guts. Note the locations of all the controls: on/off, volume, balance, tuning, and so on. Hook everything back up with jumper cables, drawing power from 2 AA batteries in a holder (Figure A). Test again.

I decided to move the listener's 2 microphones from their original position to the book's outside spine, because the spine is generally exposed whether the book is in a bookcase or just laying around.

To do this, desolder the mics from the PC board, being careful not to damage them or any nearby electronics with too much heat. Use hookup wire to extend the leads of both mics by about 6", and resolder the extended leads back to the board (Figure B).

While you're at it, extend the power leads from the transmitter and listener by about 6" as well. Protect all the new connections with heat-shrink tubing (always slip it on before soldering). Attach the power leads from the 2 components and the battery box, joining all the reds (positive) and all the blacks (negative), as in Figure C. Twist the conductors together, solder, and cover with heat-shrink.

PREPARE THE BOOK

Choose a hardcover book that's big enough in all 3 dimensions to enclose the system's components. The dust jacket should be in good condition. Pick a title that won't draw undue attention from your intended target. You don't want them to notice the book and start looking through it!

Skip a few pages and set your components centered side-by-side on a front-facing page. Draw a rectangular outline of the components with a ruler, adding ½"–1" all around, making sure there's still ½"–1" of book left around the edges. This outline will become your compartment. Cut a piece of cardboard about ¼" smaller all around, to serve as the compartment's floor.

I used an X-Acto knife and a ruler to cut the rectangle through the pages, cutting down about 1/16" each time and stopping within ¼" of the back cover. I used a binder clip to keep the cut pages out of the way while I worked downward (Figure D).

Two previous MAKE articles also describe book-hollowing techniques: "Palm Pilot Notebook" (*Volume 07, page 138*) and "Uncle Bill's Magic Tricks" (*Volume 13, page 60*).

Photography by David Simpson and Linda Kennyhertz; Sam Murphy (H)

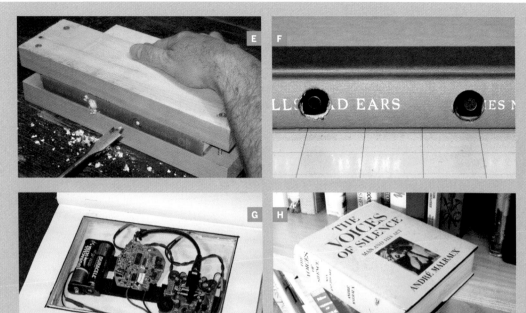

Fig. E: Drill holes through the spine for the microphones.
Fig. F: Mics and rubber dampers in place.

Fig. G: Electronics all tucked neatly inside the book.
Fig. H: Some books are not to be trusted!

Glue the cardboard floor to the bottom of the compartment and give the walls 3 coats of a 50/50 solution of white glue and water. I dried each coat for about 12 hours, with waxed paper on the top cut sheet to prevent it from gluing shut, and the whole thing under weights to keep the pages flat.

Clamp the book between 2 boards using drywall screws or handscrew clamps. For each microphone and its surrounding rubber sound damper, drill a ½" hole through the spine, clear into the compartment (Figure E). Clean away the shredded paper and paint the tunnels with the 50/50 glue solution.

HOOK UP, STRAP IN, TURN ON

I used velcro tape to mount the components to the compartment floor, and before sticking them in, I slid the microphones with their dampers into their holes in the spine. Inset the mics about ⅛" so they won't touch the dust jacket (Figure F).

Tidy the wires so that they're all well within the chamber and the book can be closed (Figure G). Test the rig again, put the dust cover on, and you're ready to save the free world!

Another cool use for this setup is as a covert 2-way communication system. Assemble 2 of these systems, then you and your fellow agent can conduct a whispered conversation in a noisy environment — listening on one another's frequency — without sitting together.

LIMITATIONS

The problem with these in-car FM transmitters is that their range is only about 15'–20'. The listener can be in an adjacent room or the room above or below if there's no metal blocking the signal.

Sorry about that, Max — you probably can't tune in from a car parked at the curb of the subject's building. For that, you might consider modifying a wireless lapel mic system, like the ones used by presenters.

David Simpson also wrote "G-Meter and Altimeter" in this volume of MAKE, as well as "Hydraulic Flight Simulator" in Volume 12 and "Explosion Engine" in Volume 13.

INVISIBLE INK PRINTER

TAKE A NEW TWIST
ON LEMON JUICE.

M016

BY MIKE GOLEMBEWSKI

JUICY SECRETS: Make your top-secret messages a mystery.

Lemon juice has been used as invisible ink for centuries. Messages written in lemon juice are invisible to the naked eye. However, when brushed with a mix of iodine and water, they become quite visible.

You can use an updated version of this technique by modifying an HP ink cartridge so that it prints in lemon juice instead of ink. Here's how to do it.

TOOLS & MATERIALS

HP inkjet printer
Color ink cartridge
C-clamp
Hacksaw
Chisel
Latex gloves
X-Acto knife
Paper towels
True Lemon crystallized lemon juice (15 packets)
Small mixing cups (2)
Wide electrical tape
Zip-lock bag
2% iodine tincture

Photograph by Garry McLeod

1. OPEN THE CARTRIDGE.

Clamp the HP ink cartridge to a workbench, with the leads facing down. Use a hacksaw to cut a ¼" groove into the cartridge, at the gap between the purple lid and the black case. Insert a chisel into the cut, and pry the lid open (Figure A). Take off the lid.

2. CLEAN OUT THE YELLOW INK.

The ink will stain skin, so wear gloves. Use an X-Acto knife to remove the yellow ink sponge from the cartridge (Figure B). Wash the sponge in a sink until the water runs clear (Figure C). Squeeze out the sponge, and put it aside.

Take a paper towel, and use it to clean the ink from the sponge holder in the cartridge. Repeat this, getting all the way to the bottom, until the towel comes out clean.

3. MAKE THE INVISIBLE INK.

Fresh-squeezed lemon juice is not strong enough to work as invisible ink in a printer, so you'll need to make a concentrated mixture. Do this by using True Lemon crystallized lemon powder. Add 4tsp of water to a mixing cup, and then pour in 15 True Lemon packets. Stir until the powder is dissolved. Soak the sponge in this mixture for 10 minutes. Make sure it's completely saturated. Push the sponge back into the cartridge (Figure D). Replace the cartridge lid and seal it shut with electrical tape.

4. PRINT THE INVISIBLE INK DOCUMENTS.

Put the modified cartridge in your printer. Create a solid yellow image at letter size, 8½×11 (Figure E). Keep printing this until no yellow ink appears on the prints. Once the pages look empty, the printer is printing only in lemon juice, and you're ready to go!

To make a secret message, print documents with solid yellow text and images (Figure F). The yellow areas will print in lemon juice, and will be invisible to the naked eye.

TIP: Once you're done using your invisible ink cartridge, wrap it in plastic and put it in a zip-lock bag; otherwise, it will dry out.

5. EXPOSE THE INVISIBLE INK!

To reveal your secret message, make a mixture of 1 part iodine tincture to 10 parts water. Brush it over the page. The message will remain white, while the rest of the page will turn pale blue!

Michael Golembewski is an artist and interaction designer who lives in Boston. For more of his work, go to mwgstudio.com.

Photography by Mike Golembewski

SPY SHRINE

VISITORS EXPLORE THE
SECRET AND EXCITING WORLD OF
INTERNATIONAL ESPIONAGE.

M016

BY LAURA COCHRANE

SMASHING, BABY: The exterior of the International Spy Museum.

The nation's only museum dedicated to the profession of espionage is located, appropriately enough, one block from FBI headquarters in Washington, D.C. With 95% of its contents devoted to real-life spies (read: not a shrine to James Bond), the International Spy Museum relies on authentic concealment devices, sabotage weapons, and cipher machines to bring to life the stories of the men and women working in this intriguing field.

Visitors start with a briefing film, followed by a tour of the School for Spies — a display of artifacts and interactive exhibits. With two floors of spy paraphernalia and lore to investigate, the museum features the largest collection of international spy artifacts on public display. When done exploring, visitors can watch a film about gathering intelligence in the 21st century.

Despite its location in a city rife with politicians, the museum tries to present a comprehensive look at espionage around the world with a focus on human intelligence, not political ideology.

For those interested in an extra dash of stealth, Operation Birthday Cake is a children's birthday party rolled into a scavenger hunt. The facilities are also available to rent out for an after-hours rendezvous — perhaps for a shaken, not stirred, martini party?

Laura Cochrane is an editorial assistant at MAKE.

Photography courtesy of the International Spy Museum

SPYING ON SPIES: Fig. A: Cold War-era KGB shoe transmitter from the former USSR. Fig. B: KGB coat with buttonhole camera from the 70s. Fig. C: U.S. WWII coal explosive device that detonates when burned in a locomotive or factory. Fig. D: A U.S. tree stump listening device used to capture Soviet radio transmissions in the 70s. Fig. E: War of the Spies exhibit, which explores Cold War-era espionage in divided Berlin. Fig. F: Operation Spy, a hands-on spy adventure challenge.

» **International Spy Museum:** spymuseum.org

Photoshop has probably influenced digital arts and crafts more than any other piece of software in the past decade. Recognizing this, this Upload section is all about imaging.

With a list price near $1,000, Photoshop CS3 is a pricey product, but there are ways around this. CS2 is selling currently for as little as $200, sealed and unused, on eBay. It should get even cheaper as CS4 hits the market, but if you want to pay even less, try version 7 for maybe $100.

Version 5.5 for Mac or 6.0 for Windows (several years old, but still with most of the features you need) can be found for as little as $50. eBay does not knowingly allow bootlegs, but make sure you ask the vendor whether the software can be, or has been, registered.

Photoshop Elements is available new for less than $100. This simpler version of Photoshop lacks some of the most powerful features, but may still suffice.

And then there's GIMP, the open source image-editing software that emulates Photoshop and costs nothing at all. Available for Windows, Linux, or Mac, it's downloadable from many shareware sites.

One way or another, powerful image-editing is now within your means. What fun things can you do with it? Here are some suggestions.

—*Charles Platt*

WARHOLIZE!

Turn your favorite blonde into a silk-screened glamour queen. By James Grant

When Andy Warhol made his famous silk-screened prints of Marilyn Monroe, he started with a simple idea: use a high-contrast black-and-white photo, and overprint it with bold swatches of color. Is that idea simple enough for us to emulate it with modern image-editing software? Let's find out. This will work in Photoshop 6 or later versions.

Figure A (opposite page) is a stock photograph that I acquired from istockphoto.com for just $14 (including the rights to reproduce it in a magazine). I chose a picture that has a plain background and is brightly lit, without much shadow. Avoid photographs that have noticeable shadows; they'll look muddy after being Warholized.

1. SAVE PHOTO AS DUPLICATE
Select the whole image, copy, and paste to create a duplicate of it in Layer 1. You'll use this version, preserving the Background layer untouched in case you need it later.

2. REMOVE COLOR
Get rid of the color in Layer 1. To do this, go to Image → Adjust → Hue/Saturation, and in the dialog box that opens, drag the Saturation slider all the way to −100.

3. MAKE TINT OVERLAY LAYERS
It'll be easier to add color overlays now, before you crank up the contrast in the photo. Create a new layer; call it "Dress," and continue by creating additional layers for hair, eyes, lips, and any other part of the picture you want to color.

4. ADD COLORS TO LAYERS
In each layer, fill an appropriate area with any color you like (you can fine-tune the colors later). Paint with a paintbrush, or fill an area that you've selected with the lasso — do whatever works for you. You don't need to be precise. Warhol slapped the color on with a "painterly" hand, which is to say, he didn't

take a lot of trouble to align the edges. After you add the layers, your picture should look something like Figure B.

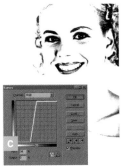

5. INCREASE CONTRAST

Now we'll increase the contrast in the photo. First click the little eyeball symbols in the Layers Palette to hide the layers you created — except for Layer 1, which must remain visible.

Click on the name you gave to Layer 1 to make it active. Open the Curves dialog box and drag 2 points to make an angular S shape, as in Figure C. The curve has been highlighted yellow, to make it more obvious. The steeper the middle section is, the more contrast you'll get. Move the anchor points left and right until you can see the main facial features in the photo, but not too much shadow. Press Enter when you're done.

6. ALLOW BLACK TO SHOW THROUGH

Now in the Layers palette, process each layer like this: click the eyeball to display the layer, click its name to make it active, and pull down the little menu near the top of the palette to change from Normal to Multiply. This will let the black parts of the image show through, just like a Warhol print where black ink showed through transparent colored ink that was laid over it (Figure D).

7. FINAL TOUCHES

Time for some cleanup. You can adjust the color edges, and maybe lighten the black areas of the lips on Layer 1. Tweak the color of any layer by using the Hue/Saturation feature. I went down to the Background layer and retrieved the pattern of the dress, to make the green area more interesting.

Does my Warholized photo stand up to a comparison with the real thing? Will yours? You'll probably find that his art was actually a little more difficult to create than it looked. And consider this: he did it the hard way, without an Undo command. In the age of Photoshop, that's something to think about.

Fig. A: A suitable photo should have few shadows. **Fig. B:** Lay on the color, without being too fussy about the edges. **Fig. C:** Use Photoshop's Curves feature so that most gray tones are changed to solid black or solid white. **Fig. D:** With each layer in Multiply mode, the black details show through. **Fig. E:** After some tweaking, you have the Warholized Marilyn effect!

James Grant creates silk-screened T-shirts in Colorado.

Create an Insect Eye

Make a multifaceted image from any portrait photo.
By Charles Platt

I wanted an insect-eye effect in Photoshop, but I wasn't satisfied with the ones that I found as plugins. So I made my own. This procedure works with Photoshop 6 or later.

1. CREATE A NEW DOCUMENT

Begin a new blank document in RGB mode, size 21"×21", 100dpi, filled with solid black.

2. MAKE A GRID

Go to Edit → Preferences → Guides and Grid to specify a gridline every ½", with 1 subdivision. Now select View → Show → Grid, and View → Snap To → Grid.

3. PASTE IN YOUR PHOTO

Open a photo, crop it to 6"×6" at 100dpi (Figure A), select all, copy, and switch to your blank document. Paste the photo, which will show in Layer 1 over the black background. Use the Move tool to drag the photo down to the bottom-right corner of the canvas.

4. DEFINE A MARQUEE

First we're going to take multiple samples from the photo at steps of ½". Choose the Rectangular Marquee tool, use Window → Show Options to display the Options Palette, and in its Style pull-down menu, select Fixed Size, then enter width 6", height 3". (Type "6 in" and "3 in" if inches are not your default units.) Feather should be 0.

5. COPY ROWS 1–6

Imagine the grid dividing your photo into 12 horizontal slices, numbered 1–12 from top to

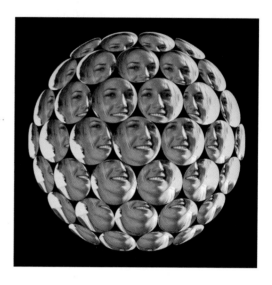

bottom. Click with the Marquee tool to select rows 1–6. Hold down Command+Option (Mac) or Ctrl+Alt (Windows) while dragging a copy all the way up to the top-right corner of the canvas.

6. COPY OTHER ROWS

Now select rows 2–7 and drag a copy of them up to touch the bottom edge of the previous copy. Repeat this procedure to copy rows 3–8, 4–9, and so on, until your copies look like Figure B.

7. COPY COLUMNS

Reset the Rectangular Marquee tool to a fixed size of 3"×21". Imagine your photo segments divided into vertical grid-columns numbered 1–12, from left to right. Use the Marquee tool to drag a copy of columns 1–6 over to the extreme left-hand edge of the canvas, then drag a copy of columns 2–7 to touch the previous copy, and so on. The result should look like an array of 49 tiles, each containing 6×6 grid squares (Figure C). Save your work so far.

Photography by Charles Platt

8. MOVE TILES INTO EYE SHAPE

Reset the Rectangular Marquee tool to 3"×3". Imagine your rows of tiles numbered 1–7, top to bottom, and columns of tiles numbered 1–7, left to right. In column 1, select and delete single tiles in rows 1, 2, 3, 5, 6, and 7. In column 2, select and delete tiles 1 and 7. In column 7, select and delete 1, 2, 6, and 7. Then select and Ctrl-drag (Windows) or Command-drag (Mac) the remaining tiles into the configuration in Figure D.

9. CHANGE INTO CIRCLES

Choose the Elliptical Marquee tool. Retain the fixed size of 3"×3". Click at the top-left corner of the first tile, shift-click the top-left corner of the next tile, and repeat on every tile until you've selected circles in all 37 of them. Then go to Select → Inverse, and delete everything except the circles.

10. SPHERIZE THE CIRCLES

Reselect the first circle and apply Filter → Distort → Spherize at 75%. Click to select the second circle and press Ctrl-F in Windows (Command-F, Mac) to repeat the same filter. After you do this to all the circles they should look like Figure E.

11. ADJUST CIRCLE POSITIONS

Use the Rectangular Marquee tool, set to 21"×3", to select row 3 of circular tiles. Drag them down until they just touch the tiles in row 4. Drag row 2 to touch row 3, then row 1 to touch row 2. Follow the opposite sequence in the lower half of the document.

12. SPHERIZE ENTIRE ARRAY

Turn off the grid. Increase the canvas size to 23"×23", then choose the Elliptical Marquee tool, hold down Shift, and stretch a circular selection area that just includes all the tiles. Apply the Spherize filter at 100% to the whole array.

GOING FURTHER

There you have it! I think 37 circles (as in the example here) is about the minimum number to get a good insect-eye effect. Naturally, if you have the patience, you can divide an image into many more.

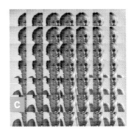
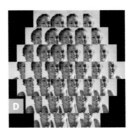
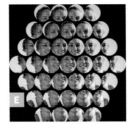

For best results, use a photo with a lot of blank space around the face.

For instance, keeping the same grid spacing as in the example above, you could begin with a photograph 7"×7" in size, pasted onto a canvas measuring 27"×27". In Step 4, your fixed size rectangular marquee would be 7"×3", and in Step 7, your marquee would be 3"×27". The rest of the procedure would remain basically the same, except that you'll need a little more patience to complete the project.

Charles Platt is the Upload section editor for MAKE.

Shadowless Closeups

A simple setup for better auction photos.
By Charles Platt

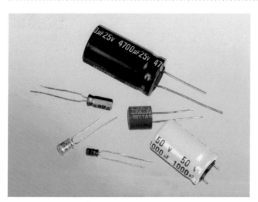

A friend of mine used to take the photographs for the sales catalogs printed by Sotheby's, the old-school auction house. Since many of the items up for sale were worth tens of thousands of dollars, they had to look good.

When I'm selling little items on eBay, I think it's still worth taking a little trouble to enhance their appearance instead of just using a flash photo of something sitting on a kitchen table. The main thing I learned from my friend is that if you place an object on an elevated glass plate and use a large, diffuse light source, your object will seem to float in space instead of sitting on its own shadow.

1. GET THE GLASS

From a hardware store, buy an 18" square of window glass or clear polycarbonate such as Lexan. Don't use acrylic, because it scratches too easily. Even polycarbonate will pick up scratches if you're not careful — which is why, personally, I prefer glass.

2. SET UP

Set up your photo area on a low table. This should be a clean area, free from intrusions from pets or family members, especially if you're using a sheet of glass, which can hurt someone badly if the person walks into it without noticing that it's there. If you don't have any photographic lights, work near a window on a day when a cloudy sky creates a neutral, diffuse glow.

3. SOFTEN GLARE

If you do have lights, you'll need to soften their glare. A reflective umbrella (about $30 from a photo supply store) is one method. A softbox is another option. This is a collapsible black fabric hood with a large, translucent white panel at the front. I use an umbrella about 4'–6' away to create sharp highlights, and a softbox up close, just 12"–18" from the object. I use incandescent bulbs because an electronic flash is expensive and unnecessary for small objects.

4. RAISE THE PANEL

Raise your glass or polycarbonate panel at least 4" above the background. I placed mine on 4 inverted plastic cups.

Photography by Charles Platt

5. CHOOSE A BACKGROUND

You can use colored paper from a craft store to complement the dominant color of your collectible, but I prefer a smooth gradation of color.

I achieved this by opening a full-page document in Photoshop and creating a gradient fill using the Gradient tool. I printed the gradient with a low-cost inkjet printer and slipped it under the sheet of glass. Although a printed gradient is less than perfect, its imperfections should disappear when it's out of focus below the glass.

6. DIAL IN COLOR BALANCE

Deactivate the flash on your camera and set the white balance to match your light source (check your camera's instructions if you don't know how). If you can get good color balance right from the start, this will save a lot of trouble later. Use the lowest ISO number to reduce sensor noise, and since this may entail a time exposure, you should use a tripod.

7. ADJUST YOUR LIGHTING

For a shiny object, sometimes you need to place a light almost in front of you, to get good "bounce" from the reflective areas. However, the light will also tend to reflect in your transparent panel.

Trial and error is the only way to achieve the best result. You may need to tilt your object relative to the camera, in which case you can tear off a small piece of duct tape, roll it into a little sticky ball, and hide it behind your object.

8. ADJUST THE APERTURE

If your camera has an aperture priority setting, choose the widest aperture. This throws the background out of focus and may also make your collectible look more 3-dimensional.

9. UPLOAD AND ADJUST

Upload your pictures into an image editing program such as Photoshop. If you see highlights on your collectible reflected in the glass or polycarbonate panel beneath it, use the Clone tool to cover them with background color.

If you want to change the background, use the

Your object will seem to float in space instead of sitting on its own shadow.

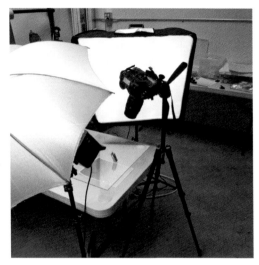

SETUP: Light from the softbox and umbrella looks yellow in this photo, because it was taken with additional daylight from a skylight.

SAMPLE PHOTO: Even a utilitarian gear-motor looks better if it's carefully lit.

Magic Wand tool to select it, then open the Hue/Saturation dialog box. You can also blur the background to hide any remaining imperfections. Finally, crop your image and adjust the resolution.

Your finished photo may not be up to the standards of a Sotheby's catalog, but compared with most pictures on eBay, it will be a work of art.

Charles Platt is the Upload section editor for MAKE.

Flipping Faces

Reveal the asymmetry in familiar features—including your own. By Erico Narita

None of us is perfectly symmetrical, but our eyes tend not to notice irregularities in faces we see many times. Anyone watching national TV news, for instance, had eight years to get familiar with Dick Cheney's lopsided grin, and a previous generation had an equal amount of time to get used to Ronald Reagan's crooked smile.

To see these features more clearly, select one half of the face and reflect it, so that the 2 sides become identical. Then select the other half and reflect that, and compare the 2 versions. The results will be surprising.

I've chosen politicians for this experiment because high-res photographs of them are available for free download with no copyright restrictions.

Of course you can create your own photographs; if you do, make sure both sides of the face are equally lit, and place your camera at least 6' from the face before zooming in to frame it. Portrait photographs should not be taken close to the subject with a wide lens, because this tends to create distortion.

When you've chosen a picture to work on, the following steps will apply to Photoshop 6 or later versions.

1. SPLIT THE FACE WITH A LINE

After opening the image, create a new layer in the Layers Palette and use the Line tool to stretch a straight line down the center of the face, from the top of the head, along the bridge of the nose, and through the center of the chin.

This may be difficult, because the chin might not line up with the nose. But don't worry if the line isn't completely vertical.

2. COPY THE LEFT SIDE

Now click the Background layer in the Layers Palette. Select the left side of the face with the Lasso tool, holding down the Alt key in Windows or the Option key on the Mac, to make it extend a straight line.

Place the line parallel to your guideline but just a little to the right of it, to allow yourself some margin for overlap. Continue around the entire left side of the photo, then copy and paste. This will copy the left side into a new layer — call it "First Layer Left."

3. CREATE A NEW LAYER

Select all of First Layer Left, then copy and paste again. This creates another new layer — I'll call it "Second Layer Left."

4. CREATE A MIRROR IMAGE

Click the Second Layer Left in the Layers Palette to make it active, then go to Edit → Transform → Flip Horizontal. This creates a mirror image.

5. ALIGN THE HALVES

Go to Edit → Transform → Rotate and turn the selection until the angle of its left edge matches the angle of the right edge of First Layer Left. Drag the image till it aligns and overlaps the one beneath.

6. CLEAN THE EDGES

Use the Eraser set to 50%, and drag it down the overlapping edge to hide the transition.

When you select half of a face and reflect it, the results can be surprising.

Fig. A: Dark lighting on one side of Vice President Dick Cheney's face creates a strange effect when it's duplicated on both sides.

Fig. B: The differences in Senator Joe Lieberman's face seem subtle — until you look more closely.

Fig. C: Late President Ronald Reagan's left and right sides almost look like 2 different people inhabiting the same body.

7. REPEAT ON THE RIGHT SIDE

Merge down Second Layer Left to First Layer Left, and you have your first reflected face. Now you can go back to your background layer and repeat the whole process, except this time select the right side of the face. Finally, get rid of the guideline that you drew at the beginning by trashing that layer.

MIRRORED IMAGE

If you want to see the asymmetry of your own face, you don't need a camera. From a hardware store, buy two 12" square mirror tiles, tape 2 edges together (so the tape forms a hinge), and angle them 90° to each other. Then look into the angle between the tiles.

The tiles flip your image twice, so that you see yourself as other people see you instead of the way you normally see yourself in a mirror. This doubles the subjective appearance of all your little asymmetries — which can be disturbing, although you should remember that other people probably don't notice the differences that seem obvious to you.

In any case, asymmetry doesn't seem to be a significant disadvantage in most occupations. It certainly didn't interfere with the ambitions of Dick Cheney or Ronald Reagan.

Erico Narita is a graphic designer in New York City.

1+2+3 Alien Projector By Brian McNamara
Shine an alien, large or small, on any wall.

This simple projector shines an image of an alien on the wall. It uses an LED as the light source and projects an image varying in size from a few inches to several feet. The simple circuit consists of only a battery, resistor, switch, and LED.

1. Build the basic circuit.
Trim the resistor and LED leads to ¼" long, and solder 1 end of the resistor onto the negative lead of the LED. Cut about 6" of wire, and solder it to the positive lead of the LED.

Put a piece of heat-shrink tubing over the black (negative) wire of the 9V battery clip, then solder the wire to the other leg of the resistor. Solder the red (positive) wire from the battery clip to the middle terminal of the switch. Place heat-shrink tubing over this wire if you wish.

Solder the positive LED wire to one of the outside pins of the switch.

2. Build the case parts.
Print the templates of the front and back plates, cut them out, and tape them to the balsa wood sheet. With a sharp knife, cut out the inside of the template first, then the outside. Use the cross on the back template to mark the hole for the switch.

Cut two 1¾" lengths of balsa wood stick. With the knife, drill a ³⁄₁₆" hole in the middle of one, and a ¼" hole in the middle of the other.

Drill a ¼" hole in the middle of the balsa back plate where it's marked, then glue the stick with the ¼" hole to the back plate.

3. Put it all together.
Glue the LED into the stick with the ³⁄₁₆" hole. Fit the switch to the back plate. Fit the LED stick into the middle of the PVC pipe, then glue the front plate onto the pipe. Snap the battery into the battery clip, then fit the back plate into place.

To set up the projector, place it in a dark room 1'–4' from the wall. Move it closer to the wall for a smaller image, and farther away for a bigger image. The best thing about this project is that you can cut out any simple image and project it onto the wall, perhaps pumpkins for Halloween and a tree for Christmas.

Brian McNamara lives near Canberra, Australia. By day he works in a university workshop designing and repairing equipment for a biological research facility; by night he designs, hacks, and bends kids' toys and musical instruments.

YOU WILL NEED

LED green
Resistor 330R
Switch SPST panel mount
6" of wire
9V battery and battery clip
1¾" PVC pipe, 6½" long
³⁄₁₆" balsa wood sheet, 6"×4"
³⁄₈"×½" balsa wood stick, 4" long

Heat-shrink tubing (optional)
Saw
Wire cutters
Wire strippers
Sharp knife
Hot glue gun
Solder
Soldering iron

Download templates from: makezine.com/16/123_alien

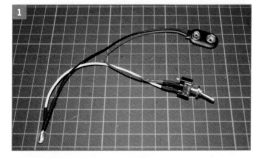

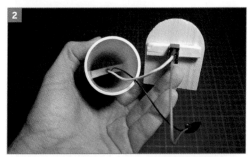

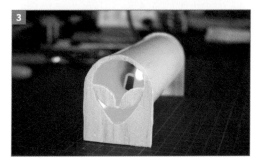

Photography by Brian McNamara

Make: Projects

Making is an art as well as a science, and these projects demonstrate a unique aesthetic. Snap an aerial masterpiece with your R/C pole photography setup, or sculpt recycled foam using a hot wire, 12 volts, and your imagination. Then capture the beauty of acoustic waves in visible form, using not much more than a vibrating platter and a coneless speaker.

Photograph by William Gurstelle and Karen Hansen

POLE'S-EYE VIEW

By William Gurstelle

A POLE-MOUNTED AERIAL PHOTOGRAPHY RIG

Sometimes nothing is as important as perspective. My goal in photography is often to find a view no one else has found, to be able to see things from unusual and insightful vantage points.

The most practical way to obtain the elusive aerial perspective is by attaching a camera to a pole. While not trivial, it's not complicated, either. Making a pole-mounted camera rig like the Sky Eye takes about a day, not including trips to the store. You can make the rig and use it the same day.

I've experimented many times with aerial photographic techniques for obtaining the much-sought-after bird's-eye view. First, there was the specially rigged kite, way back in the first issue of MAKE (*Volume 01, page 50, "Kite Aerial Photography"*). And I've dabbled with taking pictures from R/C aircraft, helium balloons, model rockets, and so forth.

While these approaches are novel, I find them limiting because the photographer is at the mercy of uncontrollable factors. There may be too much wind to loft a balloon, or not enough to fly a kite. Rough terrain or low visibility may make it impossible to launch and recover a rocket-borne camera. And none of these methods works indoors, say at a stadium. The Sky Eye works in all these conditions.

Set up: p.109   Make it: p.110   Use it: p.113

William Gurstelle is a MAKE contributing editor and is currently hard at work on *Make:* television, coming to public television in early 2009. makezine.tv

Photograph by Karen Hansen and William Gurstelle

HOW'S THE WEATHER UP THERE?

The Sky Eye design will work with most digital cameras. It doesn't require you to remove the cover of your camera or rewire or solder any circuitry. Instead, it employs one of the most useful devices in the maker's arsenal of cool tools: the radio-controlled servomotor. By building a lightweight, sturdy mounting frame, attaching your camera, and affixing a pair of servos, you can inexpensively shoot photographs that others have paid hundreds if not thousands of dollars to capture.

If you choose, you can make your shooting sessions more efficient by attaching a wireless video camera to the camera mount. Acting as a remote viewfinder, the video camera allows you to frame your shots from the ground.

It's important to remember you're making a one-of-a-kind tool based on the materials available to you and the camera you already own. So, you'll need to improvise and work out some of the details for yourself.

BUILD A SKY-HIGH EYE

1. Optional video camera
2. Digital camera
3. Shutter servo
4. Tilt servo
5. Battery pack
6. Radio control receiver
7. Painter's pole and brush extension

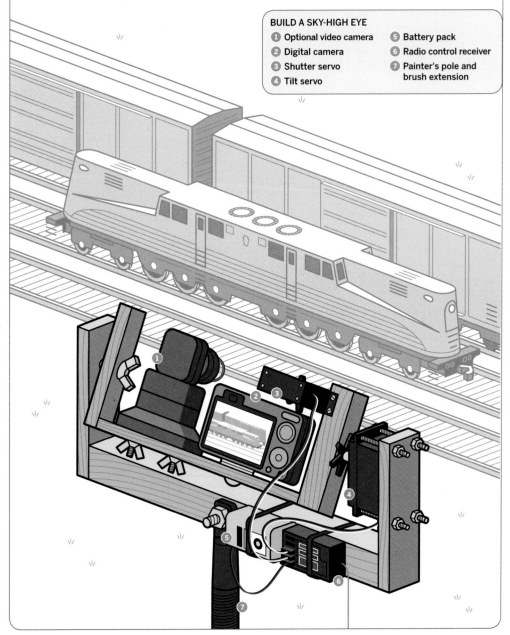

Illustration by Nik Schulz

SET UP.

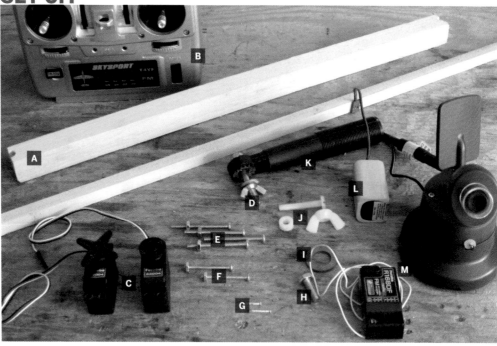

MATERIALS

[A] Wood framing The following parts worked fine for my Canon SD850, Futaba servomotors and (optional) wireless, video camera. You may have to change the dimensions slightly to fit your camera.
1"×1"×13" long
1"×1"×5" long
1"×½"×10" long
1"×½"×4½" long (3)

[B] Radio control transmitter, receiver, and battery **designed for ground use. Don't use R/C airplane or helicopter frequencies.**

[C] Hobby servomotors (2)

[D] ¼" carriage bolt, 2" long

[E] #8×2" machine screws with nuts (4) **for the upper servo**

[F] #6×1¼" machine screws with nuts (2) **for the lower servo**

[G] Small wood screws or brads **to fit the bottom servo arm**

[H] ¼"-diameter machine screw, 1" long

[I] Plastic hose washers (3–6)

[J] ¼"-diameter nylon bolt, 1½" long, with nut, ¼" nylon spacer, and 2 washers

[K] Painter's brush extension **with a socket that fits the extension pole**

[L] Battery pack (optional)

[M] Wireless video camera and receiver (optional)

[NOT SHOWN]

Digital camera

Cable ties

Telescoping painter's extension pole

#8×2" machine screws with nuts and washers (2) (optional) **to mount the video camera**

Portable television (optional)

Photograph by William Gurstelle

MAKE IT.

BUILD YOUR
SKY EYE CAMERA RIG

START ⟩⟩

Time: **Less than a Day** Complexity: **Easy**

1. MEASURE THE CAMERA AND SERVOMOTORS

The first step is to size up your camera and servos with a ruler and record the following dimensions:

A. The distance from the end of the camera to the center of the shutter button.

B. The distance from the end of the camera to the center of the tripod socket.

C. The height of the camera, from the camera bottom to the top of the shutter button.

D. If you choose to include the wireless video camera viewfinder, measure and record the overall width and height of the video camera.

E. The distance from the center of the servo arm to the servo mounting holes.

F. The distance from the center of the servo arm to its tip.

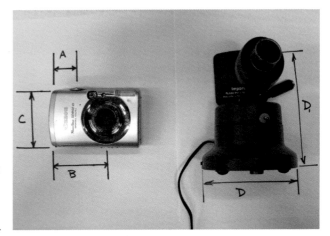

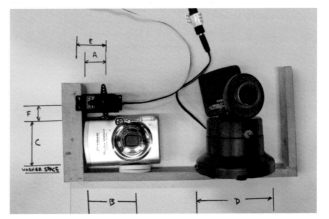

Photography by William Gurstelle

2. BUILD THE FRAMES

Using the wood framing dimensions from the Materials list, build the camera-holding top frame (shown at the top of the photo), using glue and nails. You can use wood screws, dowels, or pegs instead of nails if you prefer.

Next, build the mast attachment, or bottom, frame (shown at the bottom of the photo).

3. MOUNT THE SHUTTER SERVO AND CAMERA(S) TO THE TOP FRAME

3a. Mount the shutter servo to the left upright of the top frame (the 1"×1" piece). Position it so that its centerline is at a height of (C + F + 2 washers) above the surface of the base piece where the camera will be mounted, i.e. the combined length of the camera height (C) plus the length of the servo arm (F) plus the thickness of 2 hose washers.

Drill 2 holes in the upright with a #25 drill bit (about 5/32"), corresponding to dimension E. Attach the shutter servo to the upright with two #6 machine screws, and fasten it securely with nuts.

3b. Drill a ¼"-diameter clearance hole in the base for the camera-mounting machine screw, at a position such that the shutter button is directly underneath the servo arm when the arm is fully extended. This distance varies between cameras and depends on both dimensions A and B.

NOTE: Once you've successfully positioned the servo arm, the shutter will trip each time the servomotor rotates.

Insert the 1"-long, ¼"-diameter machine screw into the camera-mounting hole on the top frame. Place a few hose washers over this bolt before securing the camera. As you tighten the camera by turning the machine screw, the hose washers compress; this will give you a degree of fine adjustment of the shutter servo in a subsequent step.

3c. (Optional) If you're going to add the wireless video camera, drill mounting holes for it on the top frame, corresponding to dimension D, then attach it with two #8 machine screws, nuts, and lock washers. Take care to align the digital and video cameras so that they point at the same target.

4. MOUNT THE TILT SERVO AND RADIO GEAR TO THE BOTTOM FRAME

4a. Mount the tilt servo to the bottom frame, as shown here, using four 2"-long #6 machine screws. Install the nylon bolt, spacer, and nut on the bottom frame exactly opposite the tilt servo's axis of rotation. Take time to carefully align this axis so the servo can easily and smoothly control the tilt angle. Attach the tilt servo's control arm to the top frame with very small wood screws or wire brads.

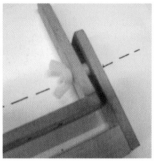

4b. Attach the paintbrush extension to the bottom frame using a ¼"-diameter, 2"-long carriage bolt.

Then attach the radio control receiver and the receiver/servo battery pack to the bottom frame using cable ties. Pull the cable ties tight so the components can't come loose.

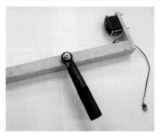

NOTE: For information on how to build the optional control panel if you're using the wireless video camera, go to makezine.com/16/polecamera.

5. MOUNT THE FRAME TO THE POLE

The last task is to screw the brush extension into the pole. The qualities that make a good photography pole are light weight, stiffness, and strength.

I used a 23' telescoping extension pole with the delightfully descriptive name of Mr. Longarm. Available in the paint department of hardware stores, it's not expensive and works adequately, although it does flex quite a bit. Fully extended in a stiff wind, you'll really feel Mr. Longarm's sway. More intrepid makers may want to experiment with chromoly steel or carbon fiber poles.

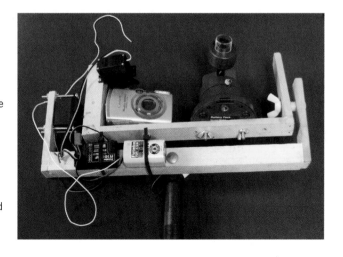

FINISH ☒

USE IT.

MAST APPEAL

SKY EYE OPERATION

1. Plug all servomotors and batteries into the slots on the radio receiver. If you don't know how to operate an R/C radio, read the manual, talk with a knowledgeable friend, or visit a model/hobby store.

2. Turn on the R/C transmitter and operate the levers to see which one controls the tilt servo and which one controls the shutter servo. Depending on your preferences, you may want to rearrange the servo plugs on the receiver. I plugged the tilt servo into radio channel 1 and the shutter servo into channel 3, thereby putting the tilt action and the shutter release on different joysticks.

3. Now you're ready to take pictures. Turn on the camera and extend the pole to a reasonable height. Raising and lowering the pole can be a bit dicey with the camera mounted to the end. Work slowly and deliberately.

4. Mast photography is best done by 2 or more people: 1 to operate the camera shutter and tilt controls and (at least) 1 assistant to hold the mast or pole. To take pictures, the assistant rotates the pole so the lens faces the desired direction, and then the camera operator adjusts the tilt servo so that from the ground it appears that the lens is pointing directly at the item to be photographed. If you've equipped your rig with the wireless video viewfinder, frame the picture using the image on the television.

The camera operator then holds down the shutter joystick until the servomotor arm depresses the camera's shutter button. If the servo fails to operate the shutter, readjust the tightness and alignment of the bolt that holds the camera against the top frame and washers.

SAFETY CAUTIONS

Tipping: The farther the pole is extended, the greater the tendency to tip and fall. Use extreme caution. Also, the greater the angle of tilt, the greater the tipping tendency. Hold the pole as close to 90° vertical as possible.

Buckling: Depending on the strength of the pole, the weight concentrated on the end of the pole could cause it to buckle or fold, especially when the pole is fully extended. The tendency to buckle becomes more pronounced as the pole is held at increasing angles from vertical. Test the strength of the pole before fully extending it or holding it at an angle.

> **⚠ CAUTION: Keep the camera and pole under control at all times, and most importantly, away from overhead wires and power lines.**

Photography by Karen Hansen and William Gurstelle

5-MINUTE
FOAM FACTORY

By Bob Knetzger

FRY BY WIRE

What keeps your coffee warm but also rides the cold Pacific surf? What's in the Rock and Roll Hall of Fame but makes an annoying, squeaky sound? Even though it's banned in over 100 cities, you can find it just about everywhere. What is it? It's expanded polystyrene (EPS) foam.

Styrofoam is a great insulator (for hot drink cups and wall insulation), lightweight and stiff, and impervious to water (great for surfboards). Unfortunately, it's also impractical to recycle and can be an unsightly part of the waste stream. Our landfills and waterways are filling up with discarded coffee cups, store meat trays, and take-out packaging.

With this easy hot-wire foam cutter, you can reuse this leftover EPS foam to create treasures from trash!

Once you've mastered the basic foam cutting techniques, go to makezine.com/16/styrocutter to learn more cool tricks. Create a double-cut, 3D teddy bear shape, spin a compound-curve cone, and cut a stack of foam sheets to make a blizzard of snowflakes.

Set up: p.117   Make it: p.118   Use it: p.121

Bob Knetzger (neotoybob@yahoo.com) is an inventor/designer with 30 years' experience making fun stuff. He's created educational software, video and board games, and all kinds of toys from high-tech electronics down to "free inside!" cereal box premiums.

Photograph by Garry McLeod

THE CUTTING EDGE

EPS is a thermoplastic foam that can be cut with a hot wire like a warm knife through butter.

You can buy expensive, commercially made hot-wire cutters, but I'll show you how to build a super-simple DIY design quickly for next to nothing, and how to get great results with some clever accessories and foolproof techniques.

For the cutting wire, this design uses a fine wire made of nichrome (nickel-chromium). It's held vertically on a table and kept taut by a bent aluminum arm. A model train transformer is used to convert AC power into a controllable and safe 12 volts DC. As the current passes through the wire, it warms up. This design lets you guide the pieces of foam into the stationary wire and slide the foam around to make effortless cuts.

HOT-WIRE CUTTER EXPOSED

This cross-section view shows how the conductive aluminum rod completes the circuit to provide current to the nichrome wire.

To make perfect cuts, use cutout cardboard shapes as templates.

1. Nichrome wire
2. Aluminum rod
3. 2×4 base
4. Wire anchor
5. Pegboard
6. Red test lead — positive
7. Green test lead — negative
8. Transformer
9. EPS foam

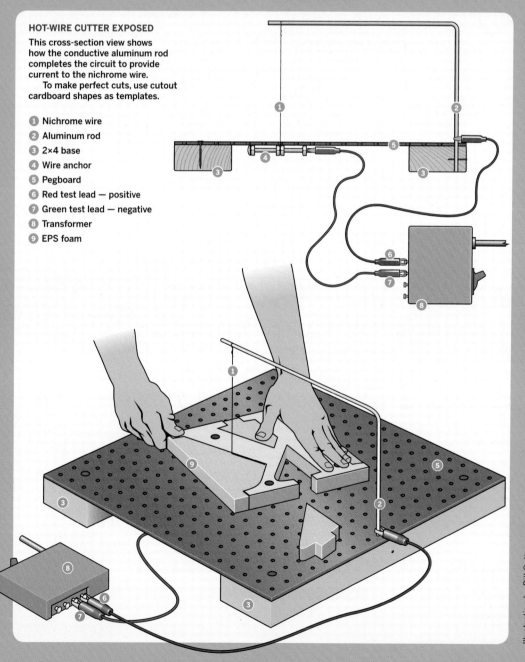

Illustration by Bill Oetinger

SET UP.

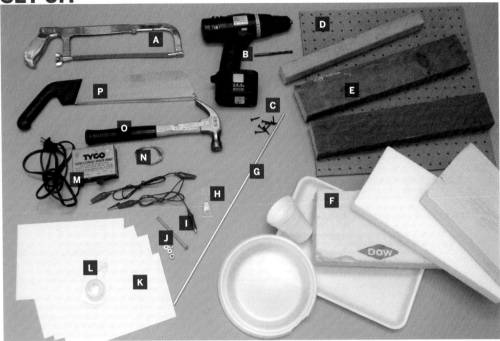

MATERIALS

Scrounge your workshop for scrap materials to build the cutter. None of the dimensions are critical, so feel free to adapt the sizes shown here to use what you've got.

[A] Hacksaw

[B] Drill and bits

[C] Screws and nails

[D] 18"×18" pegboard any thickness, tempered or not

[E] 2×4 lumber or any dimensional lumber

[F] EPS foam material Sure, you can buy it at a craft store, but why not get creative and recycle? Once you start looking, you'll find lots of EPS foam in everyday items you'd otherwise throw away:

» Grocery store meat trays in cool colors like black, blue, red, and yellow!
» Picnic plates
» Coffee cups
» Fast food containers
» Packing materials

» Styrofoam coolers
» Leftover chunks of house insulation — Dow Styrofoam "blueboard" and Owens Corning InsulPink are both extruded polystyrene (XPS), not made from little beads, and give excellent results!

[G] ¼"-diameter aluminum rod, 21" long

[H] Super glue or any cyanoacrylate glue

[I] Test leads with alligator clips (2)

[J] Bolt (1) with nuts (4) any size

[K] Paper or cardboard for making templates

[L] Transparent tape

[M] Model train transformer One with variable DC control is ideal.

[N] Nichrome wire, 0.010" diameter, with a resistance of 7Ω per foot. Get it at a hobby store or online.

[O] Hammer

[P] Wood handsaw

[NOT SHOWN]

Multimeter with Ω setting (optional)

Screwdriver

Tape measure

Photography by Bob Knetzger

MAKE IT.

BUILD YOUR STYRO CUTTER

START »

Time: 1 Hour   Complexity: Easy

1. MAKE THE CUTTER BASE

1a. Measure and cut the pegboard to make a tabletop. Mine is 18"×18". Then measure and cut 2 pieces of 2×4 to make leg rails.

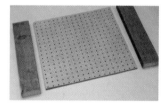

CAUTION: Wear eye protection when using all tools, and to be extra safe, wear gloves when cutting with the hot wire. You don't want to touch it — it's over 200˚F!

1b. Use screws to attach the top to the rails. Then place a ¼" drill bit through the outermost middle pegboard hole above one of the 2×4s and drill all the way through the rail. This will be the hole that the rod fits into.

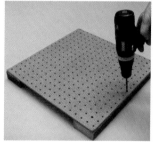

2. BEND AND INSERT THE ROD

2a. Cut the aluminum rod to 21" in length. Mark off 12" and bend the rod about 90°. After bending the rod, insert the short leg into the hole you drilled in Step 1b and align as shown here. Mark the closest pegboard hole beneath a point 1" or so from the tip of the rod. Then mark the top surface of the rod directly above the marked hole.

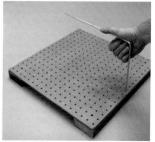

2b. Use the hacksaw to make a shallow notch across the top of the rod where it's marked.

2c. Insert the rod and drill a small ¹⁄₁₆" pilot hole through the rail and into and through the aluminum rod. Drive a nail into the hole and through the rod. This prevents the rod from swiveling in its hole as you cut in different directions.

3. MAKE AND ATTACH THE CUTTING WIRE

3a. Thread 2 nuts onto the bolt. Wrap, then tie the nichrome wire around the bolt, and add the remaining 2 nuts. Tighten the 2 nuts in the middle to pinch the wire.

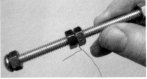

3b. Align the nuts so that they all lie flat. Put a drop of super glue on them to lock them tight.

3c. Thread the nichrome wire through the marked hole in the pegboard (from Step 2a) and pull it up vertically. Press the tip of the rod down slightly and hold it there. (You want a little springy tension to keep the wire taut.) At the same time, make a loop in the wire so that the tip of the loop just reaches the deflected rod. Hold that loop and tie it off in a simple overhand knot.

As you press the rod down, slip the loop over the rod and into the notch. Let go. The gentle spring force of the rod should make the wire taut. If it's too loose, shorten the wire by tying another knot. Trim any stray ends.

4. CHECK THE CIRCUIT

Remember your high school physics about electricity?
Voltage equals current times resistance: $V = I * R$

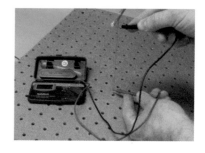

You can measure the resistance of the wire with a voltmeter:
set it to "ohms" and measure the wire by placing 1 probe at each
end point. My wire measures 7Ω. My transformer puts out 12 volts
DC. Plugging that into the formula gives:

$12 = I * 7$ or $12 / 7 = I$
so $I = 1.71$

So the current needed is just under 2 amps. My train transformer is rated at 2 amps, so that's good, at
least for short time periods. The resistance of the wire will change at various temperatures, so the current
drawn will vary. Many train transformers have a built-in thermal breaker — if they get too warm, they'll
shut off. If that happens, unplug the transformer and let it cool down. It should work again later.

5. POWER UP!

Use the alligator clip leads to attach the train transformer. First,
unplug the transformer. Connect the first lead from one side of the
regulated DC contacts to the bolt underneath the table. Connect 1
clip of the second lead to the remaining DC voltage contact on the
transformer. Make sure the variable control is at its lowest setting,
then plug in the transformer.

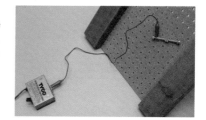

Ready to test it? Finally, connect the last alligator clip to the base
of the rod. You've created a circuit that sends current through the
wire. Adjust the transformer's control so that the wire gets warm
— not glowing red-hot. No heat? Check your connections and
make sure the clips aren't touching each other at the transformer.

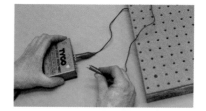

Test your cutter with a scrap of foam. Place the foam on the
table surface and gently slide it into the hot wire — it should
cut easily with just the slightest pressure. Adjust the voltage if
needed. Don't press too hard, or you'll pull the wire into an arc
and your cuts will be curved instead of straight.

Use that last connection to the rod as your on/off switch. You
can see when it's connected and that the hot wire is "on."

FINISH ⊠

NOW GO USE IT »

USE IT.

START CUTTING!

You can cut free-form shapes easily — just keep the foam moving at a smooth, constant speed. You'll notice that the slightest wiggle in your movements will result in a wavy or ridged part! Here are a few tricks for easy cutting and perfect parts.

STRAIGHT CUTS

Make a guide from a piece of 1×2 lumber and some ¼" dowels. Drill 2 holes on 1" centers and slip in pieces of dowel. Plug these dowel pins into the pegboard and you've got an adjustable fence for smooth, straight cuts.

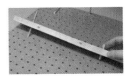

» Position the fence close to the wire to cut thin strips.
» Position the fence farther away to make wide cuts and to square up blocks.
» For size adjustments of less than 1", remove 1 of the dowel pins, and use just 1 pin as a pivot. Swivel the fence around the pin to adjust the angle until the distance between the wire and the fence is just right. Place a second long dowel pin behind the fence at the closest pegboard hole. Fast and easy!

CIRCLES

Make a circle cutter guide out of a dowel and a nail. Cut a dowel to a length so that when inserted into a pegboard hole, it's flush with the tabletop. Drill a ¹⁄₁₆" pilot hole into the end of the dowel and insert a small nail — head first. Tap it in with a hammer. Now you have a dowel with a pointy pin sticking out — be careful!

Place the dowel in any pegboard hole. The distance from the wire to the pin will be the radius of your circular cut. Energize the wire. Slide a piece of foam into the wire and impale the foam on the pin. As you spin the foam around on the pin, you'll cut a perfect circle. Turn off the wire and remove your part.

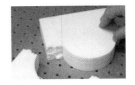

ANGLED CUTS

You're not limited to 90° cuts. Build an elevated and angled wedge to make beveled cuts! Make a wedge of the desired angle and nail it to a block with dowels on the bottom.

Plug the wedge into the pegboard and slide the foam to make your angle cut!

For more super cool techniques for cutting hyperbolic, toroidal, and conic shapes, as well as how to use templates, go to makezine.com/16/styrocutter.

CHLADNI
PLATE

By Edwin Wise

BEAUTIFUL ACOUSTICS

Use a broken speaker, bits of wire, and tape to prepare a coneless voice coil driver, then use it to generate standing waves on a sheet of metal, making sound visible. Magic!

My knowledgeable friend Robin once said that you don't need to worry about having too big an audio amplifier, because speakers are usually damaged by under-powered amps working too hard and clipping the signal, creating rough square waves with too much power. I learned that this is true when I melted a speaker's coil by running a strong 20Hz signal through it, to drive a vortex cannon (*MAKE, Volume 15, page 114*).

On the bright side, I now had a nice speaker magnet to use as the foundation for something else I wanted to try, a Chladni plate!

Early acoustics researchers Robert Hooke and Ernst Chladni (CLOD-knee) found that fine powders sprinkled on a vibrating plate would settle in patterns that showed how the plate was vibrating. They got their glass and metal plates vibrating for their experiments by running a violin bow across the edges. In our updated version, we'll generate the vibrations using a voice coil driver, which is basically a speaker without the cone.

Set up: p.125   Make it: p.126   Use it: p.131

Edwin Wise is a software engineer with 25 years' experience. He develops software during the day and explores the edges of mad science at night. He can be found at simreal.com.

Photograph by Ed Troxell

WAVE WATCH

Vibrating surfaces move and flex in some places while remaining relatively still in others. Powder sprinkled on a flat, vibrating surface will flee the moving areas and settle in the still areas. When the vibrations are steady and at the right frequency, this creates a pattern of standing waves that shows how the material is vibrating.

HOW IT WORKS

A speaker's magnet assembly has a ring magnet, a pole piece, and front and rear plates that hold them together.

The ring magnet creates the magnetic field that the voice coil pushes against.

The pole piece helps contain and shape the magnetic field.

The voice coil consists of a thin-walled former and the coil itself, the wire wound around it. Electricity moving back and forth through the wire turns the coil into an electromagnet of varying polarity and strength that moves within the ring magnet's field.

The voice coil connects to the plate.

The signal generator feeds a pure oscillating signal into the voice coil, and the resulting vibrations travel outward and bounce back from the plate edges. We adjust the signal's frequency, and when the plate can physically resonate at a particular frequency, it settles into a stable pattern: a standing wave.

The nodes are the still areas where the powder collects.

The antinodes are the clean areas where the plate moves the most.

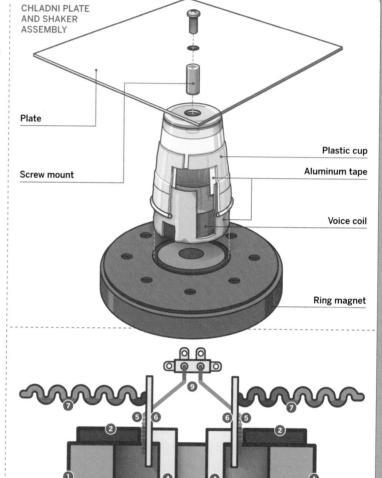

CHLADNI PLATE AND SHAKER ASSEMBLY

Plate

Screw mount

Plastic cup

Aluminum tape

Voice coil

Ring magnet

SPEAKER DRIVER

1 Ring magnet
2 Front plate
3 Rear plate
4 Pole piece
5 Voice coil
6 Former
7 Spider
8 Vent
9 Signal wires

PATTERNS

The pattern depends on the shape of the surface and the frequency of the vibrations; higher frequencies produce denser patterns. Since acoustic musical instruments produce sound by vibrating, these visualizations can be used to understand and improve an instrument's design.

Illustration by Damien Scogin

SET UP.

MATERIALS

[A] Speaker

[B] Wire-wrap or magnet wire, 30 gauge

[C] Spray paint (optional) gloss black

[D] Signal source such as a signal generator or computer running a program such as ToneGen (nch.com.au/tonegen)

[E] Amplifier such as a stereo amplifier, that the signal source can plug into

[F] Epoxy glue

[G] Aluminum tape

[H] Packing tape cheap plastic

[I] Small plastic cups (2) should be roughly the same diameter as the coil inside the speaker, which in my case was 2". Try condiment containers.

[J] Graphite powder sold in hardware stores as a dry lubricant

[K] Rigid styrofoam sheet, ½" thick, 4" square from insulation or packing material

[L] Plastic tubing, 7/32" outer diameter (OD), 5/8" long available at hobby and craft stores

[M] Metal tubing, 7/32" inner diameter (ID), 5/8" long available at hobby and craft stores

[N] Thin metal sheet, 2' square, about 22 to 24 gauge

[O] Nylon bolt, #10-24, about 1" long You'll cut it down to size.

[NOT SHOWN]

Fine powder such as fine sand, gelatin powder, salt, or sugar

Awl or nail and hammer

TOOLS

[P] Drill

[Q] Straightedge

[R] Wire cutters

[S] Sharp knife

[T] Marking pen

[U] #10-24 tap and drill bit

[V] Sandpaper

MAKE IT.

BUILD YOUR CHLADNI PLATE

START ⋙   **Time: An Afternoon**   **Complexity: Easy**

1. TEAR DOWN THE SPEAKER

The first step is to tear apart a speaker to get at the good bits. We'll need the magnet assembly and the former, which is the cylinder that the voice coil wraps around.

1a. Cut through the speaker cone around its outside edge and around the dust cap in the center. Remove the cone.

1b. Cut the outside edge of the "spider," the flexible ring-shaped fabric membrane under the cone that runs between the former and the speaker's frame.

1c. Lift the spider and former out of the ring magnet and speaker. At some point you'll need to snip the braided wire that leads into the coil.

1d. Gently untangle, extract, and unwind all the old wire from the former. Carefully prod the former into good shape. If the former has rough spots, sand them down. If it's bent, unbend it. If it's sticky, rub on graphite powder to seal it.

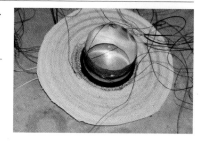

2. WIND A NEW COIL

If your speaker coil is deep enough to let you mount an extension and has wire beefy enough to handle some high-voltage driving, you can skip this step and add the riser (Step 3) to the existing coil. Otherwise it's a better bet to wind your own.

2a. Create a new former by wrapping a tight loop of packing tape around the old former with the sticky side out, so you can slip it off later. It may take several tries to get a clean, smooth cylinder.

2b. Wind a new voice coil onto the tape. Start by taping the end of your magnet wire securely to the spider on both sides, leaving some extra length. Then wind a tight, single-layer coil around the former with 50 to 100 loops. Don't make multiple layers of wire, or the coil will jam in the magnet or overheat.

TIP: When winding a coil, the first loop is the most important, because it sets the standard for the rest. To make winding easier, pull the wire off of its spool tangentially, so it doesn't kink. To help with this, I used a large Allen wrench in a vise as a spool holder. Spin the former in your hand as you pull the wire and lay it down firmly.

2c. After you're done winding, spiral the outer end of the wire down the former and tape it to the spider. Cut the wire and strip both ends.

2d. Use squares of aluminum tape to cover the coil and at least ¼" of the packing tape that sticks out underneath, but leave a small gap down 1 side so that the aluminum doesn't create a conductive ring. Make sure no tape is sticking to the original former.

TIP: Use several small squares of tape rather than 1 piece, because aluminum tape is tricky to manipulate, and a single cylinder of tape would fit poorly on the conical-shaped cup later.

2e. Slide your new coil off the former.

TIP: Calculate your coil's resistance by multiplying its length by the wire's resistance. For example, my coil had a 2" diameter, 60 windings, and 30-gauge wire whose resistance is about 0.1Ω per foot. This gives $(2 \times \pi \times 60) \times 0.1/12$ = about 3Ω.

3. ADD THE RISER

The plastic-cup riser makes a strong connection between the coil and the Chladni plate.

3a. Insert the narrow end of a plastic cup into the coil until it stops, and draw a line around it along that edge.

3b. Cut away the wide part of the cup so that it fits through the coil. Use small bits of tape to tack the cup onto the coil, with the wire coil side toward the cup and the empty tape side away from the cup.

3c. Use aluminum tape to tape the cup to the coil from the inside, covering all the wire, ¼" of the packing tape, and at least ¼" of the cup. You can add more aluminum tape to the outside of the cup for strength, but don't tape very far down past the coil or it may become too thick to slip into the speaker.

NOTE: This coil/cup connection will take a huge amount of abuse during operation, so make it strong; we don't want it to slip or buckle.

3d. Use a sharp knife to cut a clean edge through the 3 layers of tape about ⅛" beyond the coil. Remove the excess tape.

3e. Cut a ½"-thick styrofoam disk that fits snugly into the bottom of the plastic cup, and use epoxy on its face and edge to glue it in.

3f. The plate's mounting point is built from a plastic tube, tapped for a screw mount, nested inside a rigid metal tube. Cut the metal and plastic tubing to ⅝" and epoxy them together. After the epoxy sets, tap the plastic tube for #10-24 threads.

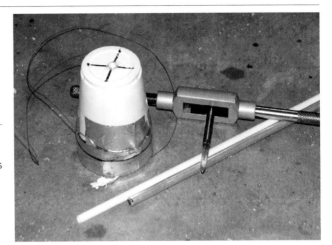

3g. Mark the center-bottom of the cup and drill a ¼" hole. Epoxy the tapped mount into the hole so that it's straight and nearly flush with the outside of the cup. Your driver assembly is complete.

4. ADD THE STOP RING

The stop ring around the voice coil holds the weight of the Chladni plate so that the coil doesn't sink too far down into the magnet.

4a. Test-fit the driver assembly into the speaker magnet. Find a reasonable position with most of the wire coil inside the magnet, and mark that position on the driver.

4b. Make a plastic ring by cutting another plastic cup ¼" down from its rounded lip.

4c. Fit the ring over the driver assembly with the cut edge pointing up toward the threaded tube, and the lip facing the magnet side and running along the marked line. Cut the ring as needed to fit, and epoxy it into place, holding it with small squares of tape as it cures.

5. PREPARE THE METAL SHEET(S)

5a. Take the metal sheet and cut a clean square about 24" on a side. Lay a straightedge across each diagonal pair of corners and draw an X through the center with a thick marker. Use a knife to make a finely scribed X in the dark ink.

5b. Use an awl or nail to dent the precise center of the X, then drill a ¼" (or so) hole through the plate.

5c. (Optional) If you have enough metal, you can mark and cut a disk, a violin shape, or any other shape.

5d. (Optional) Paint the Chladni plate black, and then un-sticky the surface with "the poor man's Teflon" by rubbing graphite powder into the dried paint.

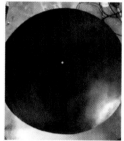

TIP: To find the center of balance of a flat shape, hold any point along the perimeter with a plumb bob hanging down, mark along the plumb line, then rotate the shape and repeat at another perimeter point. Where the lines cross is the center. You can also simply find where the shape balances flat on 1 fingertip.

5e. Using a nylon #10-24 bolt, fasten the plate to the driver.

5f. Connect your signal source to one of your amplifier's inputs and the speaker wires to an output.

5g. Finally, gently settle the driver into the magnet, down to the stop ring. That's it!

FINISH☒

NOW GO USE IT ≫

LET'S MAKE SOME NODES

OPERATING INSTRUCTIONS

1. Sprinkle some powder across the plate.
2. Starting at a signal of a few hundred hertz, slowly turn the amplifier power up until the powder starts to vibrate. Adjust the frequency and volume until patterns appear!

To reduce friction, I periodically rub graphite powder over the plate and brush off any excess. Then my other powder slides around on this slick surface like a cat in roller skates.

Clean patterns will appear for only those frequencies that resonate with the plate. On smaller plates only very high frequencies will show a stable pattern of nodes and antinodes; on larger plates, lower frequencies will resonate as well. On a large plate with a high frequency, you'll see a detailed pattern across the plate.

On a round plate, you'll mostly get concentric circles, with the number of circles indicating the ratio of the driver frequency to the plate's natural fundamental frequency. In such cases, the driver is playing a harmonic (or multiple) of the lowest frequency that the plate produces naturally when you strike it. With some frequencies, you'll see a serpentine pattern on the round plate.

On a square plate, or a plate in the shape of a violin or other complex shape, the resonances are more complex and interesting.

If you want a permanent display of your vibrational patterns, photography is the way to go. Although if you have an extremely effective filter mask and a high tolerance for a horrible mess, I would think that laser printer toner would make a nice pattern, and then, using a heat gun from below, you could fuse it to the metal for a permanent display. I haven't tried this, however. Also note that laser toner is extremely bad for your lungs, laundry, and household harmony.

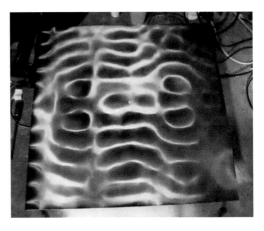

KEEP YOUR POWDER DRY

Any fine powder will work, but the finer the powder, the more sensitive it will be to vibrations, enabling it to work at lower volumes. If you're using a coarse powder, you may need to turn up the volume quite a bit before it bounces into place.

If the powder is sticky or overly fine, it may not want to bounce at all, but instead may stick to the plate and ignore even the most abusive volume levels.

Some powders, such as salt or sugar, will absorb moisture from the air and melt (especially here in Texas), making a terrible mess. Gelatin and graphite powder are both very fine, but tend to stick to the plate over time. Fine white sand should work nicely, with minimal mess.

The traditional superfine substance of choice is *Lycopodium* powder, which is the spore of a particular fern. This powder is used by magicians and pyrotechnicians as a flash powder, and it can be found at chemical supply houses.

See Edwin Wise's Chladni plate in action at makezine.com/go/chladni.

Surface Mount Soldering

Techniques for making modern circuits.

By Scott Driscoll

When cellphones were housed in briefcases, manufactured electronics had easy-to-solder leads. Now phones fit in pockets, and the smaller surface-mount devices (SMDs) inside are driving through-hole components into extinction.

SMDs can cost less than their old-school equivalents, and many newer devices, including most accelerometers, are only available in SMD format.

If you design printed circuit boards, using SMT (surface-mount technology) and putting components on both sides makes them cheaper and smaller. This may not matter on a robot, but it helps a project fit into a mint tin or hang off a kite.

SMDs are designed for precise machinery to mass-assemble onto densely packed PCBs. Their tiny leads may look impossible for human hands to work with, but there are several good, relatively inexpensive methods that don't require a $1,000-and-up professional SMT soldering station.

Photograph by Pat Molner

TOOLS

What you need depends on what you're doing and how much of it (see story).

[A] Soldering station

[B] Flux **felt pen, brush bottle, or needle bottle**

[C] Flush wire cutters

[D] Solder

[E] Lint-free wipes

[F] Hot plate **or coffee pot warmer or skillet**

[G] Embossing heat tool **from an art store**

[H] Dental picks

[I] Vacuum pickup

[J] Tweezers

[K] Hemostat

[L] Solder paste

[M] Chip Quik SMD **removal kit**

[N] PanaVise

[O] Temperature-indicating marker **or thermocouple**

[P] Toaster oven

[Q] Loupe **or lighted magnifying glass**

[R] Acid brush

[S] Desoldering braid

[T] Dry tip cleaner **or sponge**

[U] Isopropyl alcohol

[V] Stereo zoom microscope, 30x

[W] Hot air station

[NOT SHOWN]

Soldering tip

X-Acto knife

Mylar stencil

Small squeegee

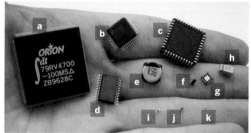

TYPICAL SMD PACKAGES

[a] QPF208
[b] QPF44
[c] PLCC
[d] SOIC
[e] Electrolytic capacitor
[f] SOT23
[g] QFN
[h] Tantalum capacitor
[i] 805 resistor
[j] 603 resistor
[k] 402 resistor

Photography by Scott Driscoll

IRONS, HOT AIR, AND TOASTER OVENS

We'll look at 3 methods of SMD soldering. The easiest components have feet or other accessible contacts that lay flat on the board's pads. These you can connect with a soldering iron. A quick touch of the tip, and a bit of solder will naturally flow under the foot and make the connection. This is the magic of SMD soldering — capillary action does most of the work for you.

Other SMD packages have their contacts on the underside, out of reach. You can solder these in 2 ways: individually, using solder or solder paste and a jet of hot air, or en masse by positioning all components on the board with solder paste between each contact and its pad, and then heating the board on a skillet or in a toaster oven to "reflow" the board (melt the paste) and make all the connections.

BASIC SMD SOLDERING

Each method has its own tools and supplies. Here are the ones you'll need for iron-soldering the simplest SMDs: resistors, capacitors, and IC (integrated circuit) packages with leads.

» Fine-tipped industrial **tweezers** let you pick up and align small components. Also helpful are **hemostats**, **dental picks** (for fixing bent leads), and an **X-Acto knife**.

» **Flux** is the secret sauce in surface-mount soldering. It removes oxides from the connections so that solder can bond to them, and also helps to distribute heat. During normal through-hole soldering, you heat the joint with an iron and then melt solder wire against it, which lets the flux in the solder's core melt out and clean the joint. With surface-mount soldering, solder is often melted *on the iron and then transferred to the joint* — a mortal sin in regular soldering. The flux tends to boil off during this transfer, so you need to add more to the connection directly. Flux comes in 3 types of container: felt pen, brush bottle, and needle bottle.

» You can solder all but the most finely pitched components using a **lighted magnifying glass**, and you can use a $10 **loupe** with 10x magnification for the finest. If you expect to do a lot of SMD work, get a **stereo zoom microscope** with up to 30x magnification (try eBay).

» I recommend getting a temperature-controlled **soldering station**, at least 50 watts, which will probably cost $50–$120. A cheap 15W iron will work on some things, but will be slower and more frustrating. A good soldering iron is especially important if you're using lead-free solder, which requires higher heat.

» Soldering stations include a sponge, but a **dry tip cleaner** lets you clean a soldering tip without lowering its temperature.

» **Soldering tip** selection is a matter of personal preference. I prefer a small 1/32" (0.8mm) chisel or screwdriver tip because it can hold a bit of solder at its end. I don't recommend tips smaller than 0.6mm, as solder tends to draw away from the point. Bevel/spade/hoof tips are designed to hold a small ball of solder at the end, which is useful for the drag-soldering technique explained later.

» Use 0.02" or 0.015" diameter, flux-cored **solder**. To get the hang of SMT, I'd recommend starting off with lead-based solder, which is slightly easier to work with.

» **Desoldering braid** or wick is a fine mesh of copper strands that you can use to remove excess solder.

» For removing SMDs without a hot air station and myriad special nozzles, use the Chip Quik **SMD removal kit** (item #SMD1, $16 from chipquik.com). The kit contains a low-melting-point metal that when mixed with existing solder causes it to remain molten for a couple of seconds — long enough to flick off the component.

» A small vise such as a **PanaVise**.

Install a 1206 Resistor

Now we're ready to install a surface-mount resistor. Note that the resistive element in an SMT resistor is exposed and colored, and it should face upward to dissipate heat. The number 1206 means that the package measures 0.12"×0.06". A 603 package is 0.06"×0.03", and so on. Let's get started.

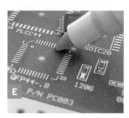

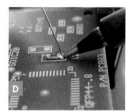

1. Add flux to the pads (Figure A). This may not be necessary for 1206s, but is helpful for 603s and 402s, where melting solder wire directly on the connection will likely deposit too much. A lightly tinned tip may provide all the solder necessary. As a rule, if you're melting solder wire directly onto a connection, you don't need additional flux, but if you're carrying solder to the joint with an iron, you do.

2. Add a small amount of solder to 1 of the 2 pads (Figure B).

3. Use tweezers to hold the 1206 in place while touching the junction between chip and pad with the iron. You should feel the chip drop into place as the solder liquefies underneath (Figure C).

4. Solder the other side by holding the iron so it touches the chip and board and adding a small amount of solder (Figure D).

Install a QFP (Quad Flat Package)
QFPs are square IC packages with leads all around. The distance between the leads, called the *pitch*, is typically 0.5mm or 0.8mm, but some are 0.4mm.

1. Flux the pads (Figure E).

2. Align the QFP over its pads with tweezers or dental picks (Figure F).

3. Add a small drop of solder to the tip of the iron. This part is key: you want a small drop to hang off the end (Figure G).

4. Tack 1 corner by sliding the tinned tip up against the toe of the lead (Figure H). The solder should quickly wick under the lead. Check alignment and tack an opposite corner. Sometimes I add more flux on top of the leads after tacking.

5. Continue touching the toes of the leads with the iron to complete the chip. You should be able to solder several leads with 1 load of solder on the tip. With practice, you can slowly drag the tip over the feet and "drag-solder" an entire row with 1 pass (Figure I).

6. Use the loupe to check for bridges and sufficient solder (Figures J and K).

7. Remove any shorted or bridged connections by touching the leads with a clean iron tip or applying solder wick (Figure L).

Alternately, there's the "flood and wick" method, which involves flooding all the leads with solder and then removing the bridges with wick. Surface tension holds some solder under the leads even after wicking. I hate to argue against something that works, but folks in the industry don't recommend this technique because it can overheat the board or component, and the wick might detach pads.

Install a PLCC (Plastic Leaded Chip Carrier)
PLCCs have legs that fold back under the package rather than sticking outward. The steps are similar to soldering a QFP: flux the pads (Figure M), align the part, tack some corners, flux some more, and solder. Keep the iron in contact long enough for the

solder to wick around the back of each pin. I like to lay a length of 0.02" solder along the pins and then press it into each pin with the iron (Figure N).

SOLDERING NO-LEAD SMDs

The following tools let you handle IC packages without leads, like QFNs (quad flat no-lead) and BGAs (ball grid array), that defy soldering with an iron.

» You can buy a **hot air station** with temperature- and flow-controlled air for under $300 from Madell (Figure O, background; madelltech.com). Instructables.com also has a wonderful array of DIY hot-air machines. If you're feeling less adventurous, a $25 arts and crafts **embossing heat tool** (Figure O, foreground) also gets the job done. Avoid ordinary heat guns; their nozzles are too big and they're too hot for SMD work.

» **Solder paste** consists of tiny solder balls floating in flux gel. It comes in 2 forms: in syringes, for applying to contacts individually, or in jars, for applying en masse with a **mylar stencil** and **squeegee** (see sidebar). Some distributors require fast shipping on solder paste, since its lifespan decreases without refrigeration.

» A **hot plate** can preheat the board to 212°F–250°F in order to limit the time and energy required when applying solder or hot air. This is optional, but it mimics the large-scale manufacturing process and reduces the risk of damaging boards or components. Preheating is especially helpful if you're using lead-free solder or if the board contains large, heat-absorbing ground planes. Preheaters are also

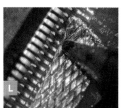

SOLDER PASTE TYPES

Solder paste comes in either syringes or jars. With a syringe, you should apply small, Hershey's Kiss-shaped drops to individual pads on the PCB, and thin lines on packages with rows of pins. I like a 22-gauge needle. In the oven, the paste will wick to the connections and avoid bridging (for the most part). Don't bother trying to put paste on every little contact individually, because it will slump (spread out) anyway when it heats. You can buy solder paste syringes from Chip Quik, Zephyrtronics, smtsolderpaste.com, and others.

Paste in jars retains its form, and you can quickly apply it to all the pads on a board using a squeegee and a laser-cut mylar stencil. Getting the right amount of paste — between having too little solder and bridging leads — takes some trial and error. For stencil material, try stencilsunlimited.com.

Both types of paste come in either "no-clean" or water-soluble formulas. With water-soluble paste, the flux residues are corrosive and must be removed.

available from Madell or Zephyrtronics (zeph.com), but a $7 Mr. Coffee hot plate works for small, single-sided boards.

» You can reflow a board in a **toaster oven**. Look for one that can heat up to 480°F (250°C) in less than 5 minutes, which will let it reflow all the solder without baking the board. Since toaster ovens don't have their 0-to-480°F speed marked on the outside of the box, I'd advise using a small one, or a large one that's more than 1,400 watts.

As an alternative, sparkfun.com has tutorials and blog entries that recommend using a skillet instead of a toaster oven for boards that carry both plastic and large metal connectors. The downside of a skillet is that it only works with 1-sided boards.

» Stencilsunlimited.com sells **temperature-indicating markers** that change color when a particular temperature is reached, to let you know when to stop applying heat. You can also monitor temperature with a **thermocouple**.

» An **acid brush**, **isopropyl alcohol**, and **lint-free wipes** clean up flux residues. I keep the alcohol in a pump bottle that dispenses as needed and prevents the rest from evaporating.

» A **vacuum pickup tool** can help place larger components that tweezers can't hold, although fingers do a decent job, too.

Install a QFN (Quad Flat No-Lead)
The recommended method with these chips is to use a stencil with solder paste, but you can also get by with regular solder and hot air.

You needn't apply solder to a chip's bottom-side heat sink, which is present on many motor amps and voltage regulators, but if you do, it shouldn't exceed 0.01" in thickness.

Also, you'll probably need to reflow it individually with a direct shot of hot air or solder it through a hole drilled underneath.

1. Flux and tin the bottom connections on the QFN (Figure P).

2. Flux and tin just the outer pads (Figure Q).

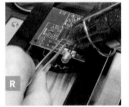

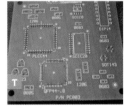

SMD PROTOTYPING

Prototyping with SMDs is more difficult than quickly plugging through-hole components into a solderless breadboard, but SchmartBoard (schmartboard.com) carries breakout boards that port any SMD to standard 0.1"-spaced through-hole pins.

For prototyping, you still have to solder the chip onto the breakout board and then remove it later to install on the final board (unless you just solder in the breakout board, which takes up space). But the breakout boards are perfect if you have a limited number of SMDs that you need to interface with through-hole components, and you aren't making your own PCB.

In my experience, it's faster to skip the breadboarding stage and go straight to a PCB prototype of the whole circuit. You can fix mistakes by scraping traces and jumpering with small, 30-gauge "green" wires. I've found that drawing schematics on a computer is more reliable than dealing with a million breadboard wires, although it's less immediate.

3. I recommend preheating, especially if you're soldering the heat sink.

4. Apply hot air about ¾" away in a circling motion until you feel the chip drop. Surface tension from the molten solder should pull the chip into alignment.

You can also nudge the chip with tweezers to make sure it's correctly seated; it should spring back into position (Figure R).

5. Check the sides with a loupe to make sure the markers line up with the pads (Figure S).

SOLDER A DOUBLE-SIDED BOARD

If the board has components on both sides, you need to use a toaster oven rather than a skillet.

1. Apply solder paste, using either a syringe or a stencil and squeegee (see "Solder Paste Types"), to whichever side of the board has lighter components (Figure T; PLCCs are the heaviest).

2. Place the components using tweezers, fingers, or a vacuum tool. It's alright if the smaller components aren't perfectly aligned; they'll snap into place during reflow (Figure U).

3. Reflow the board in the toaster oven. I use binder clips to suspend it above the rack (Figure V). Paste and component manufacturers recommended a precise 3-phase sequence:

3a. Preheat and evaporate solvents in the paste at 300°F (150°C).

3b. "Soak" between 300°F and 350°F (150°C–180°C) for 1–2 minutes to let the flux remove the oxides.

3c. Run up to about 425°F (220°C) for 1–1½ minutes to melt the solder.

What I do is simply turn the oven on max, wait for all the solder to melt, then count to 15 and open the door.

More complex boards and BGAs might require greater precision. A thermocouple or temperature-indicating marker lets you see when you've reached your target temperature.

For more control, sites like articulationllc.com and thesiliconhorizon.com sell controllers that plug into toaster ovens and let you program and run time-temperature sequences, although most toasters don't heat up fast enough to give a controller much to work with.

4. After the first side is cooled, apply solder paste, place components, and cook the other side. Surface tension will hold the lighter bottom components in place.

My results with the project photographed here were about 25 bridged connections on the 208-pin QFP with 0.5mm pitch, and a couple here and there on the other packages, but the majority turned out OK.

Scott Driscoll (scott@curiousinventor.com) is an IPC-certified soldering specialist, has master's degrees in mechanical engineering and music technology from Georgia Tech. He researches and writes how-to guides at curiousinventor.com.

Tinker. Learn. Play.
Repeat as needed. Preferably often.

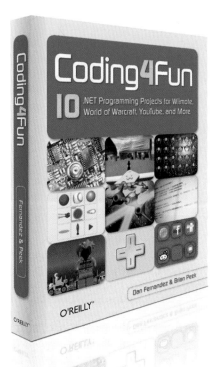

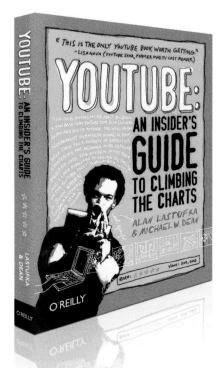

**Coding4Fun: 10 .NET Programming
Projects for Wiimote, World of Warcraft,
YouTube, and More**
by Dan Fernandez & Brian Peek
$39.99US

Build an Xbox 360 game or your own
peer-to-peer application. Get step-by-step
instructions to build ten .NET creative
projects using free Microsoft software.

**YouTube: An Insider's Guide to
Climbing the Charts**
by Alan Lastufka & Michael W. Dean
$29.99US

Learn what it takes to make a splash from
two YouTube veterans who know how to
make quality art—and have the subscribers
and millions of views to prove it.

Order now on oreilly.com

O'REILLY®

G-METER AND ALTIMETER

Double-duty aerospace instrument on a shoestring budget. By David Simpson

Here's an aerospace instrument you can build for $5 that will measure the crushing forces that a model rocket withstands and the rarified strata it attains. It isn't exactly six-sigma technology in terms of accuracy, but it's darn fun.

The device, which you install in the rocket's payload compartment, uses 2 small bands of heat-shrink tubing that slide over a dowel to record the maximum G-force and altitude attained. As the rocket accelerates, the G-force band is pushed down by washers on a spring, and as the rocket rises, the altitude band is pushed down the rod by the expansion of a pressure chamber made from a pill bottle and a rubber-balloon membrane.

The force of landing doesn't disturb the positions of the bands, which are heat-shrunk snugly over the rod and stay in place thanks to their relatively high coefficient of friction and low mass.

The "secret sauce" for both readings is the calibration step, where you mark positions on the dowel with their corresponding G-force and altitude levels. To calibrate the G-force meter, we stack increasing weights onto the spring and gauge its compression. The altimeter we calibrate using a kitchen vacuum food sealer and a commercial altimeter or barometer.

Build the Altimeter

Make the flexible membrane by cutting a 2"-diameter circle from a rubber balloon. Stretch it flat over the open end of the pill bottle, and secure it by winding button thread around several times, near the top. Tie off the thread and coat it with a thin layer of wood glue or epoxy, then trim away the excess rubber.

Cut a disk the same diameter as the pill bottle out of ¼" balsa or aircraft plywood. Drill a ¹⁄₁₆" pilot hole in the center of the disk and glue it to the bottom of

MATERIALS

Cylindrical plastic pill bottle about 2½" long to fit in a rocket payload compartment
Rubber balloon large enough to cut a 2" round piece from
Button thread or heavy-duty sewing thread
³⁄₁₆"-diameter wood dowel, 16" or longer
Model aircraft plywood, ³⁄₁₆" thick, 2" square
Balsa wood, ⅛" thick, 2" square
Balsa wood, ¼" thick, 2" square
Scrap wood board, at least 4" square for G-meter calibration stand
Small cylindrical spring, ¼" diameter
Washers: ³⁄₁₆" ID (1), ⅞" OD (2) You'll need more of each for calibration, which you can do in the hardware store aisle.
Heat-shrink tubing, ¼" diameter
Wood screw, about 1½" long
Wood glue or epoxy
CA (cyanoacrylate) adhesive gel
High-tack, double-sided foam tape
Paper Graph paper is best.

TOOLS

Model rocket with transparent plastic payload section at least 5" long I used the Estes HiJax (EST 2105), which has a 6" clear payload compartment. Or you could make your own from a spare body tube and nose cone mated to a matching "booster."
Drill and drill bits: ¹⁄₁₆", ³⁄₁₆"
Saw for cutting wood dowel
Tool for cutting 1" rounds out of thin plywood You can use a hole saw, but I rough-cut with a saw and used a Dremel tool with the 401 mandrel bit to turn the pieces against a sandpaper surface.
Altimeter or barometer a dedicated device, or you can use a hiker's wristwatch or a weather station, such as one from RadioShack
Vacuum chamber or a vacuum food sealer and vacuum bag
2" PVC pipe, 12" long
Colored pencils or fine markers

the bottle. This base will later accept a wood screw to hold the chamber fast to the tube coupler at the bottom of the payload section (Figures B and G).

Build the G-Meter

Cut a disk of ³⁄₁₆" aircraft plywood just slightly shy of the inner diameter of the rocket's payload tube, and drill a ³⁄₁₆" hole in the center.

I did this in a fun way: first I traced the circumference of the body tube onto the plywood, rough-cut

the disk, and drilled a ¹⁄₁₆" hole in the approximate center by eye. Then I threaded the disk onto a Dremel 401 mandrel bit (a shank with a screw-like head at the end) and used the Dremel as a sort of mini reverse-lathe, turning the disk down to diameter against a piece of sandpaper tacked to a board (Figure C). The mandrel bit is designed to hold polishing bits, but hey, here's a new use! After sanding down the disk, I redrilled its center hole out to ³⁄₁₆".

Glue a 3½" length of ³⁄₁₆" dowel into the disk, and mark a scale on the dowel from the base to the end, in ¹⁄₁₆" increments, alternating between contrasting colors to make it easier to read (Figure D).

Slip a ¼" length of heat-shrink tubing over the dowel, and shrink it down until it stays in place even if you shake it like a thermometer. This is the G-force band. Slide the spring over the dowel, followed by the smaller washer and then the 2 larger washers. Now shrink another ¼" length of heat-shrink — this is the altitude band.

Prepare the Adjacent Rocket Sections

Cut 2 more disks out of ⅛" balsa, to fit the rocket's inside diameter (Figure E). Drill one with a ⅛" hole in the center and use CA gel adhesive to attach it flat into the tube coupler piece that will take the bottom of the rocket's payload tube, about ½" down from the top edge. This disk will anchor the altimeter chamber.

Drill a ³⁄₁₆" hole in the center of the other disk and use CA adhesive to glue it into the nose cone, flush with its bottom edge.

The G-meter dowel will run through this disk and slide farther in when higher altitudes cause the pressure chamber membrane to balloon upward. The altitude heat-shrink band will meanwhile stay in place against the disk.

Assembly: No Bouncing Around!

Use a short wood screw to secure the payload tube coupler to the altimeter base (Figure F). Attach the G-meter base to the altimeter membrane with a ⅛"–³⁄₁₆" square of high-tack, double-sided foam tape. Use a thin film of epoxy to attach the G-meter base to the spring, the spring to the washers, and the washers to each other (Figure G).

Drill a ¹⁄₁₆" hole through the side of the payload section to allow external air pressure to reach the altimeter.

Photography by David Simpson and Linda Kennyhertz

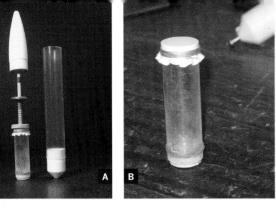
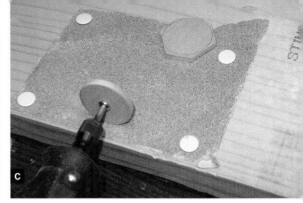

Fig. A: The combo G-meter/altimeter before installation in the payload compartment. Fig. B: The completed altimeter chamber before assembly. Fig. C: Turning wood disks to size with a rotary tool.

Fig. D: Mark the scale on the G-Meter dowel with contrasting colors for easy reading. Fig. E: Wooden disks for the tube coupler below the payload and the nose cone above.

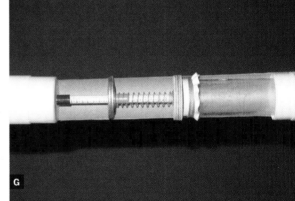

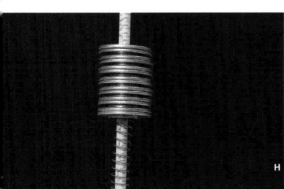
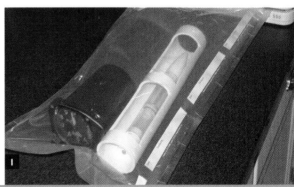

Fig. F: The tube coupler screws to the wooden base of the altimeter chamber. Fig. G: The G-meter/altimeter combo installed without the tube, to show position.

Fig. H: Calibrate the G-meter by stacking washers in multiples of carry load. Fig. I: Calibrate the altimeter alongside a working barometer or altimeter in a vacuum.

Calibrate the G-Meter

Calibrating this combo meter is more complicated than constructing it, but it's both cool and educational. To calibrate the G-meter, you first build a calibration rig. Glue a 12" length of ³⁄₁₆" dowel vertically in a wooden base and mark it the same way you did the G-meter dowel.

Slide the G-meter spring over the dowel and add the payload quantity of 1 small washer and 2 large washers. Check the reading on the scale. That's 1g, our baseline. Make a table that translates millimeter readings on the dowel to corresponding G-forces. Record the 1g reading, then add 1 more small washer and 2 more large ones, and record the position on the scale for 2g. Continue doing this until you've got readings for 40g (Figure H). Unless you need a lot of washers, you might opt to conduct this procedure in the aisle of the hardware store.

Calibrating the Altimeter

Calibrating the altimeter is even cooler, but more involved. You'll need access to a vacuum chamber; if you can't gain access to a scientific or industrial vacuum chamber, borrow (or buy) a vacuum food sealer. You'll also need an altimeter or barometer; I experimented with a RadioShack mini digital weather station, which included a barometer, but it was a challenge getting it to take instantaneous readings. I was able to do it by interrupting and re-energizing the circuit at the moment I wanted the reading. But I wound up using a good old airplane altimeter, which was simpler.

If you're using a vacuum food sealer, put the G-meter/altimeter in a cage made by cutting two 5" windows in a 12" piece of 2" PVC pipe. The cage prevents the bag from collapsing around the instrument. Put the reference altimeter or barometer and our instrument in the vacuum bag or chamber (Figure I).

Blip the vacuum button to reduce the pressure in small increments, and record the readings from the dowel and the commercial barometer or altimeter. Make an association table between the two, like with the G-meter. If you're using a barometer, you can convert pressure to altitude using the calculator at csgnetwork.com/pressurealtcalc.html.

Prepare for Liftoff

Make sure both heat-shrink bands are in their starting positions before each flight: the G-meter band at the top of the spring, just under the washers, and the altimeter band at the top of the dowel, so that it will be flush with the nose cone disk when you slide it on. If the nose cone is a little loose, secure it with masking tape. As with other model rockets, friction-fit the payload atop the booster, connect the elastic shock cord, and attach the parachute — or else you'll be digging your combo meter out of a hole!

Recovery, Reading, Reset, and Repeat

After "3, 2, 1, liftoff!" and *whoosh!*, recover your vehicle and peek inside. You may need to loosen the bottom wood screw in order to read the bands, then use your tables to translate their positions into max-g and max-altitude. You've got your max-g and max-altitude recorded on the combo!

A variation: Instead of marking millimeters on the dowels, inscribe the target units — *g*s and feet or meters — directly during calibration. Just take care not to bump anything while making the marks.

You'll find that different rockets, engines, and gross weights all affect the readings. You may need to add or remove washers and recalibrate to change the overall range of the G-meter; more powerful engines need fewer washers to register a reading while weaker engines need more.

Safe rocketeering!

David Simpson is a private pilot and teaches aviation to teenage cadets as aerospace education officer for the Civil Air Patrol in New Jersey. He can be reached at dsimpson@hydroflightsim.net.

The Key Hole
My first job was as a bicycle mechanic at a shop where we also cut keys. One fine day, a fellow came in to have his house key duplicated, and he asked me to drill a nice big hole in the key so it would be easy to find in the dark. It wasn't long after that when I drilled my own key! My current key ring has, among others, three identically shaped Schlage keys, so I cut a little off the wings of another to make it easy to identify. Try this trick — I'll bet you never go back.

—*Frank Ford*, frets.com

Find more tools-n-tips at makezine.com/tnt.

TWINKLE TOES

Add a flashing, tricolor UFO to your roller skates. By Dan Bassak

Photography by Brent Eckart

Skatetown is our local roller rink in Bloomsburg, Pa. For years, it's also been my artistic LED experiment laboratory. I started by taping single-color LEDs to skates. They were so popular that one session on the floor looked like a swarm of fireflies.

My latest LED invention is the UFO Toe Stop. It uses a disk-shaped "flying saucer" unit that contains a red, green, and blue (RGB) LED. I wanted to blend the colors in interesting ways, so I wired it to a PIC-controlled RGB microcontroller that comes with programmed routines such as fixed colors, fast-changing rainbows, or dazzling strobe effects.

I also used some translucent jam plugs (small bolt covers used in skate dancing) that I bought in the rink's pro shop as light diffusers.

1. Unscrew the boot and the old toe stop (or jam plug) from the skate plate (Figure A, following page).

Place the skate plate on the ⅛" plastic sheet (Figure B). Mark the 4 boot holes on the sheet and drill them to the same size as the boot screws.

2. Temporarily bolt the skate plate to the plastic sheet with 1" screws. Trace the outline of the plate using a sharp point and then tap the skate plate with a mallet to leave marks where the plate's protruding rivets are located. (In Step 4, you'll drill holes through these marks to allow clearance for the rivets.)

3. Mark the center of the toe stop, looking down through its threaded hole. This will be the center of your LED. Because most toe stops are at an angle, center the hole slightly toward the front of the skate. This slight angle will give higher light output from your LED toe stop.

MATERIALS (For 1 skate)

PIC 12F629 microcontroller chip
$7 from Big Clive's Shop, bigclive.com.
IC socket, 8-pin DIP All Electronics part #ICS-8,
allelectronics.com
UFO LED, RGB 3x 1-watt $10, Quickar Electronics
part #mhrgbufo, quickar.com
Resistors: ¼ watt 1kΩ (3), ½ watt 47Ω (3)
5V, 1A voltage regulator such as an LM7805,
from Jameco Electronics, jameco.com
2N7000 N-channel MOSFET transistors (3)
from Jameco Electronics
22µf 25V electrolytic capacitors (2)
Circuit board 1"×1" All Electronics #PC-1

NOTE: Alternatively, you can buy an RGB board
kit for $25 from bigclive.com that includes
everything above except the UFO LED and the
47Ω resistors. I built my own because the kit's
printed circuit board was too big for my skates.

Push-button switches, normally open (2)
All Electronics #MPB-1
Sub-mini on/off switch All Electronics #SMTS-4
9V alkaline battery
9V battery clip with 3" leads All Electronics #BST-3
Plastic project box I used a 1"×2"×3" box,
All Electronics #TB-1.
12" lengths of fine wire (4) to power the LED.
I used old printer wire.
12" length of ⅛" heat-shrink tubing
Semi-transparent jam plugs (2) $5/pair, color
GLOW, from newskates.com
New skate boot bolts (8) $5 from newskates.com
1"×10-32 machine screws with nuts (4) from any
hardware store
12" length of velcro tape
12"×12"×⅛" sheet of Delrin plastic $14,
McMaster-Carr part #8575K113, mcmaster.com.
This will build 2 insulator plates.

TOOLS

Scroll saw
Electric drill or drill press
Forstner drill bits: ⅞" and ⅜"
Assorted small drill bits
Wire stripper or small bolt cutters
Wire cutter
Long-nose pliers
Screwdrivers
Soldering iron
Solder and solder wick
Center punch
Electrical tape
#220 sandpaper
File for trimming plastic
Hammer
Rubber mallet
Skate wrenches (optional)

4. Remove the temporary bolts from the skate plate and find your center mark for the UFO LED. Drill a flat-bottomed hole (not a through-hole) with a ⅞" Forstner bit. The goal is to leave a 1/32" layer of plastic at the bottom of the hole to serve as an insulator between the metal plate and the LED. In the center of this hole, drill a ⅜" hole through the plastic sheet, to allow the UFO LED lens to poke through. Make a small notch in the rear part of the thin plastic insulator (to make it easier to run the small wires for the LED). Drill out clearance holes for any protruding rivets at this time.

5. Using a scroll saw, cut out the outline of the skate plate on the plastic sheet. Stay on the outside of the trace line. Smooth the edges; sandpaper works well. Verify the accuracy of the plastic plate against the metal skate plate; it should fit flat. Note the square notch for wires in the plastic plate (Figure C).

6. Solder 4 wires to the UFO LED, using very thin wires about 12" long that will fit inside ⅛" heat-shrink tubing after it's shrunk to 1/16". I used red, green, and blue to indicate the LED colors and white as the common wire (Figure D). Wire from an old printer cable worked well for me.

My skate plate had a gap in the metal at the back of the toe stop. Enlarge it just a little with a ⅛" drill bit to make a place to run the LED controller wires (Figure E). If your plate uses a jam nut to lock the toe stop, you may need to drill a ⅛" hole for the wires.

7. Place the plastic plate over the metal skate plate and line up the holes. Insert the LED wires through the hole or slot in the rear of the toe stop and gently place the UFO LED into the hole, making sure that the shrink tubing protects the small wires from chafing (Figure F). Place some velcro tape (the soft, loop side of the hook-and-loop pair) over the LED to cushion it. Reattach the boot by installing the plastic spacer between the boot and the metal skate plate using new bolts.

8. Drill holes in the project box for the on/off and program switches. The switches should be placed where they're accessible but somewhat protected from breakage. Solder wires to the switches, then install them (Figure G). Mount the control box to the skate plate using velcro or screws (Figure H).

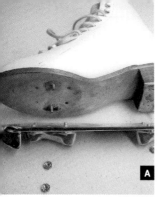

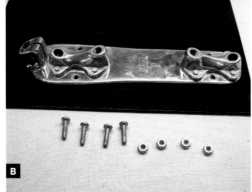

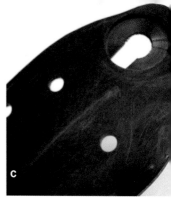

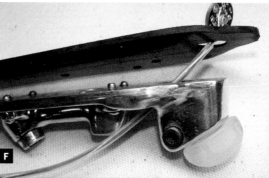

Fig. A: Remove the toe stop and boot from the skate plate. Fig. B: Trace an outline of the plate on a plastic sheet and mark the location of protruding rivets and the toe stop hole. Fig. C: Note the square notch.

Fig. D: The UFO LED unit. Fig. E: Drill a hole in the plate's toe-stop-tightening slit so you can route the UFO LED cable through it. Fig. F: Make sure that the shrink tubing protects the small wires from chafing.

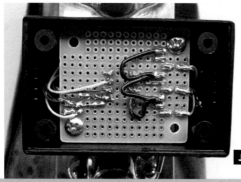

Fig. G: Drill holes in the project box for the switches, solder wires to the switches, then install them.
Fig. H: Mount the control box to the skate plate using velcro or screws. Fig. I: Install the battery between the boot and plate. Fig. J: Solder the UFO LED wires to the components on the circuit board. Fig. K: Place the cover on the control box and reassemble the roller skates by attaching the skate trucks and wheels.

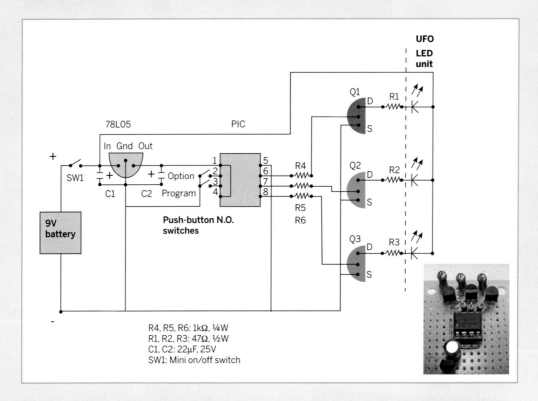

UFO
LED
unit

78L05 PIC

Q1
D R1

Q2
D R2

Q3
D R3

In Gnd Out

+

SW1

C1 C2 Option Program

1
2
3
4

5
6
7
8

R4
R5
R6

9V
battery

Push-button N.O.
switches

R4, R5, R6: 1kΩ, ¼W
R1, R2, R3: 47Ω, ½W
C1, C2: 22μF, 25V
SW1: Mini on/off switch

Plan a good route for the LED wires, from the toe stop to the control box. Drill a hole in the box for the LED wires, leaving them about 3" long.

9. Build the control board. I used a small printed circuit board that came with my plastic project box. First, I installed a socket for the PIC, and then I built the circuit around it. Make sure the PIC chip power is run through the voltage regulator and only gets 5 volts. The solder connections were placed on the outside of the PC board to make hooking up the power and LED wires easier.

I connected 47Ω current-limiting resistors to the LEDs to prolong the battery life. This was bright enough in a well-lit rink. Bigclive.com has complete kits with the circuit boards if you want, but the board was just a little too big to fit my skates.

10. Find a place for the battery. I installed the battery next to the heel of the boot between the boot and plate (Figure I, previous page). I used velcro to hold it in place. Now, plan where you'll run the battery wires, then drill a small hole in the control box for them. Insert the wire into the box.

11. Solder the battery power wires to the power

switch and the control PCB. Next, wire in the program and option push-button switches; they are pins 2 and 3 on the PIC chip (the other side of the switches are grounded). Next, solder the common (white) wire for the UFO LED to the battery positive and then the red, green, and blue LED wires to the 47Ω current-limiting resistors from the field-effect transistor (FET) outputs (Figure J).

12. Install the translucent jam plug. Turn on the UFO controller with the power switch. The LED will start displaying colors. By holding down both the program and option switches simultaneously for a short time, you can toggle the PIC between the standard and FX modes. The option switch will allow you to slow down or speed up the program mode. Place the cover on the control box and reassemble the roller skates by attaching the skate trucks and wheels (Figure J).

That's it! Now you're all set to take your skates to the rink and trip — er — *roll* the lights fantastic!

When Dan Bassak isn't experimenting with LEDs or dodging rink vipers, you can find him tinkering in the kitchen or tending the organic vegetable garden. dan@mcnelis.com

REMOTE VOLUME KNOB

Turn the knob without leaving the couch.
By Paulo Rebordão

There's nothing like hard work to get your mind running. Having to get up from the couch to adjust the volume on my old stereo amp several times each night got me thinking. I didn't want to buy a new stereo, so I came up with a plug-in remote volume controller that you can easily adapt to anything with a reasonable-sized volume knob.

My circuit uses a Picaxe-08M microcontroller, which is programmable in BASIC, and a TSOP2238 infrared receiver, which sees signals from any Sony (or compatible) TV remote. The removable device hangs off the volume knob, attached with velcro.

At the top, an R/C servomotor turns the knob through about 180° of travel, which should be enough range for most environments. At the bottom, 3 AA batteries act as a "keel," weighing that end down so that the knob is forced to turn.

Design and Program

Before actually building, I conceptualized the circuit, drew a schematic, and wrote a draft version of the software. You can download my schematic and final program at makezine.com/16/diycircuits_volume.

I prototyped the circuit on a solderless breadboard, adding the extra circuitry required to program the Picaxe-08M (as described on page 22 of the online Picaxe Manual), including a Picaxe download cable connected to my PC's serial port. From this setup I tested and tuned the software until everything worked right. I downloaded each revision to the chip by pressing F5 in the Picaxe Programming Editor.

From your remote, the software recognizes the Up/Down volume buttons as well as Mute. Pressing Mute again will restore the previous volume level. There's no auto power off, although standby

MATERIALS

Resistors: 330Ω (2), 10kΩ, 4.7kΩ
Capacitors: 100µF, 4.7µF
Picaxe-08M microcontroller
TSOP2238 infrared (IR) receiver
1N4001 diode
Analog R/C servomotor such as SuperTec S7.5
3x AA battery holder or 4x AA with 1 slot shorted
IC socket, 8-pin DIL for the microcontroller
Small toggle switch, SPST I used a push button.
Perf board, about 4cm×4cm
Male R/C servo plug
Small plastic or metal box
Small screws (2–4) for holding the servo
Sony-compatible TV remote
Hookup wire, duct tape, and velcro tape (10cm–15cm)

TOOLS

Solderless breadboard
Picaxe Programming Editor from picaxe.co.uk
Picaxe download cable
Resistors: 10kΩ, 22kΩ in circuit with download cable
Drill and small drill bits
Soldering iron and solder
Wire cutters

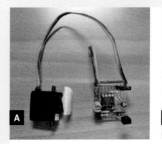

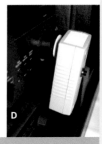

Fig. A: Circuit on small piece of perf board.
Fig. B: Project box to hold circuit and servomotor.
Fig. C: Inside the project box. Fig. D: Ready for service.

consumption is tiny.

Servomotors vary in range of movement, so the software defines 2 constants (TOP and BOT) for the motor's upper and lower limits of travel. If you use the software with a servo other than the S7.5, you'll have to tweak those numbers.

Build the Device

With everything working, I transferred the circuit to a piece of perf board, leaving out the Picaxe programming circuitry. I soldered in an 8-pin IC socket to hold the Picaxe, and plugged it in with the final version of the software already loaded (Figure A).

I selected the smallest project box where everything would fit, and then drilled 5 holes: 3 for the servo axle and fixing screws in back, and 2 on the front for the IR sensor opening and the power switch (Figure B). Screw the servo to your enclosure firmly and position the batteries as low as possible at the other end. I wedged the battery holder between 2 of the box's screw posts, and used duct tape to attach the circuit board to the inside of the box (Figure C).

I put the longest arm available onto the servo motor axle, to maximize contact with the knob. Then I covered the arm with the loops side (softer) of some velcro tape, and affixed the hooks side (scratchier)

to the front of the volume knob (Figure D).

When you first start the circuit, the servo travels to its lowest setting. This lets you match knob and servo positions by turning the volume knob to the lowest setting and then joining the velcro. It also prevents your amp from exploring the upper reaches of its volume range. Your hearing will thank you.

Test the Up, Down, and Mute buttons. Check that the movement range is adequate and, more important, that the amp's knob isn't hitting its off-stop on the way down. If so, you might have to reposition the controller slightly. Happy listening!

Improvements and Alternatives

» Replace the IR receiver for another one compatible with your remote brand. Seven different variants (TSOP22xx) cater to most brands out there.
» Include an auto power-off feature. This is desirable, but not easy to implement in a simple design.

⊞ For project software and a schematic, visit makezine.com/16/diycircuits_ volume.

Paulo Rebordão is a software practitioner but is considering moving into hardware for a firmer grasp on reality.

THE DISEMBODIED VOICE OF JUDY GARLAND SPEAKS!

How to make a Ghost Phone.
By Greg MacLaurin

Photography by Greg MacLaurin

Currently, I'm obsessed with analog telephones. I don't know why. My last obsession was with the severed hands of mummies, but let's not get into that. Today it's phones. And these Ghost Phones are fun. The idea is simple: hide an MP3 player and its headphone inside an old analog telephone, and you can listen to someone talking to you!

But before we start unscrewing things willy-nilly, let's take a moment to find out about analog phones and how they work. (I hate reinventing the wheel. I'm all into research.) Privateline.com has a good history of telephone technology, and HowStuff-Works has a technical overview at howstuffworks. com/telephone.htm. Remember, these are just starting points. Please do your own research; you'll learn something amazing in the process.

Now that you've become a font of information on analog telephones, you're ready to modify one. Here are my instructions for making a Judy Garland Ghost Phone.

1. Record audio and load the MP3 player.

1a. Decide which ghost you want to summon. Think of who the ghost is, and who they're talking to. Set a scene. All my Ghost Phones are first-person: the ghost totally monopolizes the conversation, and never lets you get a word in edgewise. Even though the ghost might ask a question, they either answer it for you or they just continue talking.

MATERIALS

Rotary telephone
Speaker from an MP3 player headphone with cord
MP3 player playing a 30-minute, custom-edited
   monologue
Female insulated spade lugs (2) for earpiece
Male spade lugs (2) for headphone jack
Electrical tape to insulate connectors
Tape to cover the hole in the bottom of the phone
Audio files Counterpoint-music.com sells a
   2-CD set of Judy's self-recorded notes for her
   unwritten autobiography.

TOOLS

Screwdriver
Wire cutter/stripper/spade lug crimper all in one!
Saw, Dremel, or metal shears
First aid kit

PROPS (OPTIONAL)

Telephone table white metal with custom,
   garishly upholstered padded seat
Vodka bottles (3) 1qt Gordon's and 1pt
   Seagram's (2)
Tonic water can
Drinking glass, 8oz with garish gold and
   black pattern
Glass pill bottles (10) with custom labels in
   3 types:
      Los Angeles (Mayer Drug Co.) (2)
      New York (Luft Drug Store) (2)
      London (Palladium Drug Co.) (6)
Large scarf in garish black and colored print

1b. Obtain or record your audio. The audio should be longer than your audience's attention span. The Judy phone has 30 minutes of audio. You don't want your audience to hear a repeat; that's just sloppy dream sharing. Also, it might be nice to filter the audio so it has the same limited frequency range of a telephone, but it's up to you.

1c. Get a small, cheap MP3 player, one that's simple, tiny, and has less than 1GB (Figure A). Find one at a thrift store or on eBay. Avoid proprietary devices, like Sony's or Apple's, that force you to download software to use the device.

1d. Load the MP3 file onto the player (it helps if the only file on the player is your Ghost Phone audio), set the player to permanent loop/repeat, press play, and you're set.

2. Replace the speaker in the earpiece.

You don't want to use the existing telephone speaker because speaker technology has advanced considerably over the past 10 or so years. Also, the analog telephone speaker's impedance doesn't match an MP3 player's.

2a. After you remove the telephone's earpiece, you'll see that the speaker is screwed into the wiring by 2 spade lugs. Remove the screws and toss the speaker.

2b. Take the headphones that work with your MP3 player and break off 1 earpiece, keeping about 3" of the wire pair. Strip ½" of insulation from the end of each wire.

2c. Crimp 2 insulated female spade lugs onto the speaker wires (Figure B), then plug the male and female spade lugs together (Figure C). No bare wires should be visible. Check for shorts. Wrap it up with electrical tape.

3. Connect the audio plug to the telephone's terminal strip.

3a. Take what remains of the broken headphones, and cut off the wires about 12" above the plug.

As you can see, the sleeve for the headphone wires contains 3 conductors: left, right, and common. You need to join the left and right wires.

To find out which is which, plug the audio plug into the MP3 player and join 2 of the 3 wires together. Take these 2 conductors and touch them to the speaker. If it plays, then permanently twist the wires together. If not, untwist and try another combination. Still not working? Untwist and try the third possible combination.

3b. Once you've figured out the wiring, crimp on male spade lugs.

3c. Take the cover off the phone and look for the main terminal strip. It's under the handset cradle. I'm pointing to it with my favorite yellow screwdriver (Figure D).

NOTE: All phones are not the same, so some experimentation may be in order.

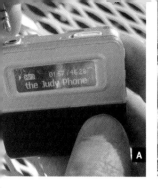

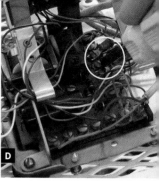

Fig. A: Copy your audio file to a small, cheap MP3 player. Fig. B: Crimp 2 insulated female spade lugs onto the speaker wires. Fig. C: Plug the male and female spade lugs together. Fig. D: Take the cover off the phone and look for the main terminal strip under the handset cradle. Fig. E: Cut a hole in the metal base of the phone so that the MP3 player can be hidden inside. Fig F: Create a mysterious scene.

3d. To add realism, we want the voice to cut off when the handset is placed in the cradle. Trace the 4 wires from the handset (2 for the earpiece, 2 for the mouthpiece), down through the cradle switch and on to the main terminal strip.

You can use either a multimeter or trial and error: turn the MP3 player on, touch your 2 MP3 male lugs to the 4 handset wires on the main terminal strip, and listen for which 2 wires on the terminal strip are for the earpiece. Screw the lugs into the terminal strip.

3e. All that remains is to cut a hole (with a saw, Dremel, or metal shears) in the metal base of the phone so that the MP3 player can be hidden inside (Figure E). File down the sharp edges and cover the opening with duct tape.

You're done! The MP3 player should play through the earpiece when the handset's lifted from the cradle.

Yes, the MP3 player must run constantly. It might be possible to connect a switch so that the player turns on when the handset is lifted from the cradle, or perhaps you can give the phone a remote switch. How about hooking up a motion sensor so that it automatically turns on when someone gets within proximity?

Another thing I like is to take the existing wall cord from the phone and rip and shred it about 12" from the phone itself. It makes it all a bit more mysterious. And I am all into mystery.

4. Create a scene.

Finally, since I'm into creating mysterious environments and making art projects far more interactive, I try to create scenes for the phones: a specific table that the phone rests on, appropriate props around it. Things that enhance and deepen the story of the ghost talking to you on the phone.

It's more than just a phone with a dinky MP3 player inside. It's a strange dream, shared. And it's wonderful that I can share this one with you!

Greg MacLaurin (gregagogo.com) is an artist and concept designer in Los Angeles, whose work for Walt Disney Imagineering, Universal Creative, and other theme park design companies is challenging and fun, but is usually so secret he can never talk about it.

CHATTER TELEPHONE

Surprise! A classic pull-toy phone that really works. By Frank E. Yost

I remember making pretend phone calls on my Fisher-Price Chatter Telephone when I was 7 or 8, and wondering if it was possible to turn it into a real phone. That question stayed with me, and when I saw a Chatter Telephone and a Crosley Princess Telephone recently at Target, I knew the answer was yes. I brought them home and made it work, and it was easier than I expected.

Disassembling the Chatter and Crosley phones was easy with screwdrivers (Figures A and B). The Chatter's dial pops off when you tap out the pin underneath with a hammer and nail. To clear room inside the Chatter, I used a Dremel and X-Acto knife to shave the bell and clicker mounts off the inside of the bottom cover. Examining the phone's workings, I saw that there were 6 elements I needed to fit into the Chatter. Here's how I handled each.

Push-Button Dial

I used a paper template to mark and cut a 3" hole in the Chatter, centered over the dial's sticker. The rest of the sticker I peeled off. I temporarily taped the Crosley dial in place in the 3" hole, turned it all upside down, and glued the dial in around its circumference. For reinforcement (optional), I Dremeled off the part of the Crosley's shell that held the dial in back, filed its edges, and screwed it back on using the original screws (Figure C, page 156).

Later, I had to grind down the bottom inside edge of the dial, to give the Chatter's eyes room to bob up and down. As a finishing touch, I removed the Fisher-Price sticker under a hair dryer, and affixed it to the new dial.

RJ11 Jacks

With a knife, I cut holes for the jacks in the Chatter's base just next to the Fisher-Price logos on the right

Photography by Frank E. Yost

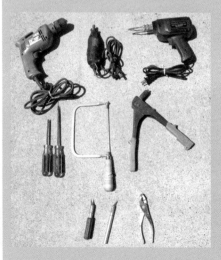

side and the back. Then I glued the jacks in place
from the inside.

Hang-Up Hinge

The Crosley's hang-up hinge is too long to tuck
inside the Chatter, so I trimmed 1⅛" off each end of
the top piece and ¼" each from the bottom. I measured and drilled two 7⁄16" holes in the Chatter for the
clear plastic plungers to stick up through (Figure D,
next page). I then made a bracket out of sheet metal
to hold the hinge underneath; download the pattern
at makezine.com/16/diytelephony_chatter.

With the plungers through the new holes, I positioned the hinge and bracket so that the plungers
moved up and down freely, and then I marked the
bracket's position, drilled four ¼" holes through the
sides of the phone, and pop-riveted it in place.

Circuit Board

I made brackets for the circuit board out of
aluminum angle; see makezine.com/16/
diytelephony_chatter. I insulated them with
electrical tape, screwed them to the board with
4 plastic nuts and bolts, and then drilled and pop-
riveted them to the Chatter through the flat part
of the base (Figure E).

Fig. C: For reinforcement, use part of the Crosley's shell and screw it back on using the original screws. Fig. D: Cut holes in the Chatter phone's cradle so the plungers can move up and down freely. Fig. E: Make

brackets for the circuit board, screw them to the board, and then pop-rivet them to the Chatter phone. Fig. F: Put the bell on the outside of the Chatter phone by pop-riveting its mount in the back.

Bell

I put the bell on the outside of the Chatter phone by pop-riveting its mount across the sound vents in the back. The 2 wires powering the bell tucked neatly through one of the vents (Figure F).

Handset

I assembled the handset last, after testing the modded body with a working donor handset. The Crosley's handset had delicate wiring that melted under a soldering iron, so I used an older phone from a thrift store. I gutted the handset, cutting the wires to the microphone and speaker. Then I used a coping saw to cut off the Chatter handset's caps, ½" from each end.

To add sufficient weight to push down the plungers, I hammered a 4" length of ¼" brass pipe into shape in a vise, threaded the curved pipe through the handle, and glued it in place. I cut out a hole for the jack, then fished the wiring through the pipe.

I drilled ¹⁄₁₆" sound holes through the end caps, following the toy's existing dimple pattern, then glued in the microphone and speaker. I resoldered the wire connections, insulated them with tape, and glued the jack in place. Finally, I taped the caps back onto the handset tightly, using precisely cut

red electrical tape that matched the toy almost perfectly (Figure F).

Conclusion

That's how I turned a classic toy into a working telephone. Now call someone! With a phone like this, you'll have plenty to talk about.

Frank E. Yost is an amateur artist who lives in Andover, Minn. He wrote the Retro R/C Racer project in MAKE, Volume 11.

Coloring Glue
When you're about to glue up a crack repair or any other job where the glue will have to fill some gaps and be visible, don't forget to add some color to the glue. It's always better to have the glue line looking a bit darker than the surrounding wood, and the closer you match the color the better. Regular powdered artist's pigments work well with most any glue, whether water soluble or catalyzed. —Frank Ford, frets.com

Find more tools-n-tips at makezine.com/tnt.

CHUMBY PHONE

A hackable platform disguised as a red hotline. By Daniel Gentleman

There's no simple explanation of the Chumby. It's an alarm clock on steroids, a digital photo album, a tiny Linux box, an internet radio player, and more. Owners can set up a queue of personalized software widgets through which the Chumby continually cycles. These widgets, made interactive through the Chumby's touchscreen and motion sensors, include news, weather, email notifications, Flickr feeds, Facebook friend status, and even Netflix queue status.

Above all, Chumby's open design welcomes hacking and crafting. The creators of Chumby offer not only their entire base of source code, but the schematics to their hardware as well, at chumby.com/developers/hardware (login required). With its "open everything" nature, the Chumby is easy to craft into a personal information appliance that can fit into any setting. In my case, this happens to be my computer desk, where I thought a bright

red, 1980s-style telephone would look perfect.

Here's how I repackaged my Chumby into an old desk phone, swapping its touchscreen for the phone's original keypad, fitting its stereo speakers into each end of the phone's handset, and putting its ports on the back, replacing the original RJ14 jack.

Chumby Inside Tour

The first step in crafting with the Chumby is learning the parts and how they all fit together. The leather case, slightly larger than a softball, has a 3" touchscreen face. The rear of the unit has a plastic assembly that houses a couple of USB ports, a headphone jack, and a pair of rear-facing speakers. On top of the Chumby, a single button under the leather toggles the control panel on and off.

Now for the inside tour. Open the unit by running a hobby knife or thin screwdriver between the

Fig. A: Chumby dissected: the original leather case, the core unit, the rear unit with Chumbilical attached, and the cotton stuffing. Fig. B: The core unit with the USB riser card in place. Fig. C: The riser circuit board (lifted here) is a primarily blank board used to lift the third component, the USB wi-fi module, off the main circuitry. Fig. D: The interior of the phone prior to gutting all the components.

MATERIALS

Chumby
Desk phone
Workable material to make a small bracket **Any kind of wood, metal, or plastic that you can size and shape will do. I used a spare, bendable metal belt from a VEX Robotics kit.**
Mounting tape
26-pin 0.1"-spaced dual-row female connectors (2) **from an electronics parts store**
26-wire, 0.1"-spaced ribbon cable, 6" long **You can use wider cable; I bought 32-wire cable and peeled away the excess.**
Small panel-mount momentary push-button switch
Insulated hookup wire

TOOLS

Hobby knife
Thin screwdriver
Jigsaw or Dremel with cutting bits **I used a Dremel, but jigsaw cuts would have been straighter.**
Cardboard
Pencil or thin marker
Multimeter

plastic bezel and the leather case on the outside to release the glue. Peel back the case from the bezel, and the main assembly, known as the "core unit," pops out. This core unit is attached to a 26-pin ribbon cable (the "Chumbilical"), so be gentle when removing it.

Disconnect the Chumbilical from the core unit and remove all the stuffing inside, so you can access the rest of the components more easily (Figure A). The other end of this ribbon cable attaches to a small "daughtercard," which carries the charger port, USB ports, headphone jack, on/off button connector, and a hidden 9-volt battery connector. Unscrew the 4 small screws that attach this daughtercard to the plastic at the rear of the Chumby, and remove the entire rear assembly.

The core unit (Figure B) consists of 3 circuit boards. The main circuit board affixed to the LCD contains all the major electronics, including the CPU, memory, display controller, and a USB port that connects to the second board: the USB riser. The riser is a primarily blank board used to lift the third component, the USB wi-fi module, off the main circuitry. The USB riser's empty back makes it a good surface for mounting additional hardware (Figure C).

E

F

G

H

Fig. E: Make a mounting bracket for the Chumby's display from a thin strap of metal. Fig. F: Cut a notch in the phone's metal baseplate to allow access to the Chumby's recessed power button. Fig. G: The Chumby's speakers fit nicely at each end of the handset. Fig. H: All done. It's fun to have a big, retro, important-looking red phone that runs Linux, tunes internet radio, and displays web information widgets.

Disassemble the Phone

Disassembling the donor phone was initially exciting but then somewhat heartbreaking, because these classic phones are such works of art on the inside.

After debating ways that I might attempt to reuse some of its components, I eventually decided to just gut the entire thing and only use its chassis, to make installation easier (Figure D).

Mount the Display

To hold the Chumby's display up where the phone's original keypad was located, I created a mounting bracket from a thin strap of metal, which I screwed into the phone's bottom plate and bent up to the proper height and angle (Figure E).

I first made a cardboard template by tracing lines around the keypad hole in the faceplate. Then I attached the metal band to the base with some of the phone's original screws and, using the template as a guide, bent the band around to hold the Chumby's touchscreen up where the original keypad sat. I affixed the touchscreen to its bracket with mounting tape, and fit the rest of the core unit underneath.

I used a Dremel with a cutting bit to cut a window for the Chumby's screen in the phone's faceplate. This is the most visible modification you'll make, so

take extra care in measuring, cutting, and finishing the square. I made a mistake when cutting one corner, and looking back, I wish I had used a jigsaw, but I used the blue "Chumby Charm" logo to successfully cover up the small misstep.

Extend the Chumbilical

The original Chumbilical is only a few inches long, and after mounting the display I realized I'd have to lengthen it if I wanted to put the Chumby's ports on the back of the phone. Initially, I hoped that I could simply strip some of the ribbon cable's wires, then solder leads connecting them to a new power plug, the USB riser, speakers, and switches. But from reading the schematics on the Chumby developer site, I learned that the daughtercard doesn't just handle ports and power, it also carries a 3-axis motion detector that sends its readings to the main board so that software can react when the Chumby is tipped or shaken. It made more sense to extend everything.

To create a new Chumbilical, I went to my local electronics parts store and picked up a longer ribbon cable and 2 female 26-pin connectors, all for under $4. I crimped a connector to each end of the cable, making sure I had the header's plastic notch and the red "pin 1" marking on the correct side each time. After closing the connectors down onto the ribbon cable, it's a good idea to test for connectivity and short circuits using a multimeter and a needle. As a final step, fire up the Chumby and test the Chumbilical to make sure all functions still work.

Remove the Speakers and Daughtercard

Once the Chumbilical worked, I set forth dissecting the rear assembly of the Chumby. The daughtercard is only a few centimeters across and is affixed to the rear assembly with several screws.

The control panel button, 9V battery connector, and speaker wires are attached to the daughterboard with plastic connectors and headers, so I gently pulled them off; no desoldering necessary. Disconnecting these made the daughtercard slide easily out of the plastic assembly.

I measured the space on the daughtercard dedicated to the Chumby's ports, and cut a matching hole in the back of the phone around its original RJ14 jack. Then I used mounting tape to affix the daughtercard to the inside of the phone case, with its ports and jacks facing out the hole.

The metal baseplate of the phone curves up around its perimeter, and its back edge covered the Chumby's recessed power button, located at the bottom of its daughter card. I cut a notch in the metal to allow access to the button (Figure F, previous page). This lets you turn its power off and on with a pen or other small pointed object, which is the way you'd do it anyway; the Chumby is meant to be left on continuously.

Move the Speakers to the Phone Headset

Using a small screwdriver, I easily broke the brittle adhesive that attached the Chumby's speakers to the rear assembly. To put them in the phone's handset, I simply soldered some extension wires insulated with heat-shrink tubing to the RJ14 jacks at each end of the existing coiled phone cable. The cable has 4 wires, so it can drive 2 speakers.

The Chumby's speaker wires are thick and easy to strip and solder, making them a good choice for hacking. Remembering to maintain the polarity, I soldered the pin connectors on the daughtercard for the speakers to the RJ14 jack contacts for the coil cable at the phone end. Then I unscrewed the microphone and speaker ends of the handset and simply lifted the 2 disk-shaped components out.

The Chumby's speakers fit nicely at each end of the handset (Figure G). I wired them to the RJ14 jack on the bottom of the handset, and stuffed cotton around each speaker to prevent them from rattling against the plastic.

Mount the Control Panel Switch

The final step was to replace the control panel switch, the squeeze switch at the top of the Chumby. The original part is a limit switch with a long, curved arm. It fits well within the Chumby, but its large, exposed contacts would be easy to short-circuit, making it not very good for other projects.

The good news is that it's simple to replace with any other type of intermittent contact switch. Initially, I wanted to set up my Chumby so the control panel switch would be triggered by the rocker that rotated when the phone was lifted from the hook. But my engineering skill failed me on this, forcing me to concede to a standard panel-mount pushbutton drilled in behind the handset cradle. Soldering the wires was easy, but now I have to disconnect that button whenever I open the phone chassis.

Future Improvements

While it's not perfect, it is fun to have a big, retro, important-looking red phone that happens to run Linux, tune internet radio, and display widgets (Figure H). Since the Chumby's schematics and source code are all available, all 3 of the following improvements should be relatively easy work.

» Build an iPod dock into the back of the unit. Chumby can read from a USB-connected iPod (but not iPhone or iPod Touch), and it has a nice touchscreen interface for playing through the speakers.

» Improve access to the power button — likely by splicing out the Chumbilical's leads for that contact.

» Connect a switch and hack software to activate the phone's "hook" buttons. I eventually want the device to launch the internet radio application when you take the handset off the hook. Then, imagine hanging up the red phone and having internet radio automatically turn off. Sweet!

Daniel Gentleman (dan@thoughtfix.com), better known as ThoughtFix, operates two blogs about mobile technology and portable Linux devices.

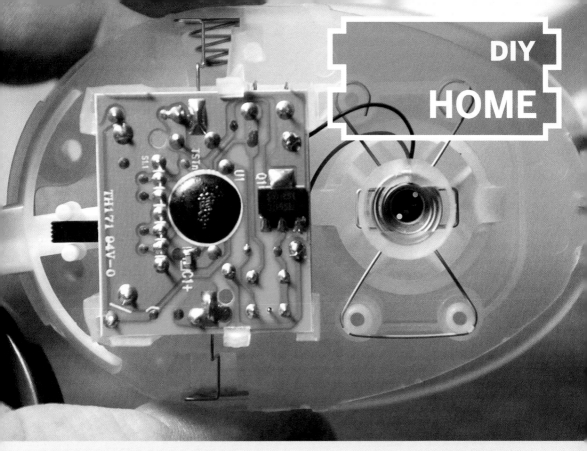

HACKING THE GLADE WISP

Make your own scent output peripheral from a piezo air freshener. By Wayne Holder

Not long ago, my 11-year-old daughter Belle wanted to create a gadget that would amuse her dog Panda by dispensing different scents for him to sniff. I had no idea how to control the dispensing of fragrances, so we took a trip to the local pharmacy and checked out the electric air fresheners.

Most of them diffused fragrances with heat or fans, but one, the Glade Wisp, claimed to use a microchip to "automatically puff" scented oils into the air. Intrigued, I bought one to see what made it tick.

The Glade Wisp runs off a single AA battery, which powers a vibrating piezoelectric disc that atomizes and disperses aromatic oil in short, smoke-like puffs. The Wisp turns out to be easy to hack — for less than $10 you can make a computer-controlled aromatic atomizer for all sorts of practical and artistic projects.

Here's how I modified a Wisp to be controlled by an Arduino board running just a few lines of code.

Before you go tearing the Wisp apart, get familiar with how the manufacturer intended it to work. Screw in the scent bottle, remove the red blocking tab from the battery, and watch the unit in operation for a few minutes.

Set the Wisp to its strongest setting, use a desk lamp to illuminate it from the side, and hold a dark piece of paper behind it; you should be able to see white vapor puff out from the top about every 10 to 15 seconds. This will show you what to expect from your modified unit. If you're curious about how the Wisp works, you can read the patent online; see the references for this article at makezine.com/16/ diyhome_aroma.

1. Hack in and hook up.

Now let's see what's inside. First, pry off the top shell, starting from the end opposite the adjustment

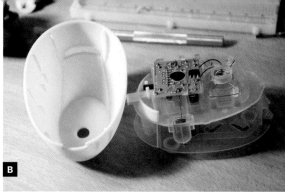

A B

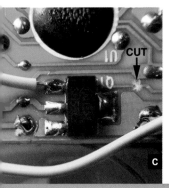

CUT

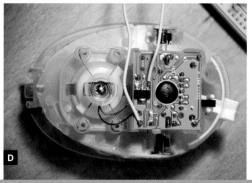

C D E

REWIRING THE WISP: Fig. A: The Glade Wisp Scented Oil Fragrancer. Fig. B: The cover removed, exposing the PC board and scent bottle. Fig. C: The power transistor with its original control trace cut and new control wires attached. Fig. D: The hacked Wisp with new control wires. Fig. E: The Wisp connected to the Arduino controller on a solderless breadboard.

MATERIALS

Glade Wisp Scented Oil Fragrancer There are a few models; I used the one shown here, not the one shaped like a bottom-heavy donut or the more expensive "flameless candle" with the flickering LED.

Arduino microcontroller board Any type that's based on the Atmel ATmega168 chip running at 16MHz will do; I used an Arduino Nano.

24-gauge hookup wire

Essential oil, ½oz, fragrance of your choice The possibilities are vast; Google "fragrance oils" or investigate candle making and soap making suppliers.

TOOLS

Computer running Arduino software from arduino.cc/en/main/software
Sharp X-Acto knife
Soldering iron and solder
Wire cutters and strippers
Flat-blade screwdriver
Jeweler's flat-blade screwdriver or drill and small drill bits
Rubbing alcohol
Glue gun and hot glue
Helping hands tool (optional) with clips and a magnifying glass

switch. The cover is held to the base with snap-fit plastic tabs, so you'll need to get a bit physical and sort of "unzip" the tabs toward the switch end (Figure B). I used a small flat-blade screwdriver.

The base and shell are rather flexible and all the circuit components are attached loosely, so I don't think you can damage the workings unless you slip and rip out one of the wires that attaches the circuit board to the piezoelectric atomizer next to it.

Inspect the PC board and locate a black, rectangular component with 3 leads coming out of one side and 1 big lead on the opposite side. It will probably be marked "3055L." This is a power MOSFET, a type of transistor, and it drives a transformer on the underside of the board, which in turn boosts the battery voltage up to the level needed to power the piezoelectric atomizer disc.

The first step in the mod is to use an X-Acto knife to cut the trace that connects the MOSFET to the control chip that's hidden under the big, round blob of epoxy in the middle of the board.

Severing this connection lets us take over control and vibrate the disc ourselves. With the blob oriented above, I severed the trace just to the upper right of the MOSFET's big lead (Figure C).

Cut 2 lengths of 24-gauge hookup wire long enough

to reach from the Wisp to your Arduino board. Connect one wire to the upper left MOSFET lead, and connect the other to the board contact nearest the lower right corner (Figures C and D). Take care not to create short circuits by bridging the MOSFET's pins, and inspect your work carefully. The MOSFET is small, so you may need a "helping hands" tool to assist.

Connect the Arduino's digital output pin D2 to the wire connected to the MOSFET, and connect the Arduino's ground pin to the other lead from the Wisp board (Figure E). That's it for the hardware mods needed to control the Wisp.

2. Program the Arduino.

I used an oscilloscope to probe the Wisp onboard controller's output, and found that it drives the atomization process by generating a 150MHz signal that lasts about 10 milliseconds. We'll program our Arduino board to mimic this signal. The following Arduino code simulates the Wisp's signal pattern, except that it tells the piezoelectric disc to "puff" every 2 seconds instead of every 10–15 seconds.

Reattach the fragrance bottle to the Wisp, then upload and run the code, and if you've wired everything correctly, you should see white puffs coming out of the atomizer immediately.

```
void setup () {
 DDRD = 0xFF;
}

void atomize (char pins) {
 unsigned int ii;
 char kk;
 while (digitalRead(8) == HIGH)
 ;
 for (ii = 0; ii < 2000; ii++) {
 PORTD |= pins;
 for (kk = 0; kk < 12; kk++)
 ;
 PORTD &= !pins;
 for (kk = 0; kk < 12; kk++)
 ;
 }
}

void loop() {
 atomize(0x04);
 delay(2000);
}
```

In the code, the **DDRD** and **PORTD** keywords configure the Arduino's D0–D7 pins as outputs to be controlled directly, by using the Arduino's port manipulation commands. The nested loops in the **atomize()** function toggle the D2 pin (specified by passing in the value 0x04) on and off a total of 2,000 times, with a very short pause after each change.

The values I chose for the delay loops make the Arduino's output roughly match the Wisp controller's, but with a shorter, 2-second delay between puffs. You may have seen other code that controls the Arduino's digital outputs by calling **digitalWrite()**, but this would be too slow to generate a 150MHz signal.

You can experiment with setting the loops to count up to values other than 2,000 and 12 to see how this changes the atomization process, but note that shorter delay times may not give the circuit's 3,300µF capacitor enough time to fully recharge between puffs, which will result in significantly decreased vapor output.

3. Substitute your own fragrance.

As interesting as it might be to have a computer-controlled air freshener (not very, IMO), the real fun begins when you replace the unit's original fragrance with something more meaningful or exotic, such as, say, the smell of freshly baked cinnamon buns, or hazelnut coffee.

To do this, you need about ½oz of an essential oil in the fragrance or aroma of your choice. The design of this bottle makes it difficult to remove the cap, but it can be done. I used a small, flat-blade jeweler's screwdriver to pry around the cap, pulling it away from the neck and breaking the glue that attached it. You can also drill a small hole in the top of the bottle.

Whichever method you use, you should clean the bottle with rubbing alcohol and let it dry, to remove as much of the old scent as possible. Once you've added your own fragrance, reattach the cap to the bottle with hot glue so it will twist back into place in the Wisp.

4. Further development: Create an aroma orchestra.

Because the Arduino's **PORTD** value lets you write to all of its digital outputs at the same time, one Arduino can control up to 6 Wisps simultaneously. You simply connect each Wisp to a different output pin and pass different values into the **atomize()** function. Using the Arduino programming environment's Serial Monitor feature, you can send keyboard

characters to the Arduino, which lets you create an instrument that plays fragrances, live. Just make sure to avoid using pins D0 and D1, which share their function with the serial port.

For example, the following code reads an input character and uses it to select which of 4 different Wisps to puff. Typing the 2 key commands the Wisp that's connected to pin D2, while typing 3 commands pin D3, and so on.

```
void setup () {
DDRD = 0xFF;
Serial.begin(9600);
}

void atomize (char pins) {
unsigned int ii;
char kk;
for (ii = 0; ii < 2000; ii++) {
PORTD |= pins;
for (kk = 0; kk < 12; kk++)
;
PORTD &= !pins;
for (kk = 0; kk < 12; kk++)
;
}
}

void loop() {
char cc = Serial.read();
switch (cc) {
case '2':
atomize(0x04);
break;
case '3':
atomize(0x08);
break;
case '4':
atomize(0x10);
break;
case '5':
atomize(0x20);
break;
}
}
```

To set this up, load the code to the Arduino board, then click the Serial Monitor button, which is the rightmost button at the top of the Arduino's development environment.

Olfactory Displays

Scentovision (1939) movie theater scents introduced at New York World's Fair
Aroma-Rama (1959) movie theater scents from *Behind the Great Wall*
Smell-O-Vision (1960) movie theater scents from *Scents of Mystery*
Odorama (1981) movie scratch-and-sniff cards from John Waters' *Polyester*
iSmell (2001) computer peripheral developed by DigiScents, never marketed
Scent Dome (2004) computer peripheral from TriSenx, trisenx.com
Fragrance Communication System (2005) networked aroma emitters and services for homes, hotels, and theaters, in development at NTT Communications, ntt.com
CineScent (2006) movie theater scents, in development, cinescent.com

This will display a set of controls near the bottom of the window. Select 9,600 baud, then type a number (2–5) into the text box and press Send. This should trigger the corresponding pin, and the Wisp it's connected to.

By using a variation of this code, you can program a set of Wisps to play your own home-theater version of John Waters' notorious Odorama, triggering a "smelltrack" that's synchronized with pictures or scenes of different aromatic subjects appearing onscreen. A small fan that blows the aromas toward the audience might be helpful here.

With some additional work, you can create more advanced scent-enabled applications, such as a network-controlled aroma generator that could receive fragrance-based "mood" messages. It's all up to your imagination and ingenuity.

Meanwhile, Belle and I will be busy working on the device she's designing for Panda, but we hope you'll have fun with this idea, too!

Find the project code and other resources at makezine.com/16/diyhome_aroma.

Wayne Holder has a classical education in computers, tinkering, and building.

USB MOTION DETECTOR

 Turn your PC into an ambush multimedia presenter. By Ken Delahoussaye

Gone are the days when people's interest could be held by simple radio or television. Today we're bombarded with information and we crave interactive experiences that don't waste a single second of our time. Advertisers recognize the difficulty of presenting messages that cut through the clutter, and they've come up with creative ways to capture our attention.

One example: the multimedia kiosk, now common in shopping malls, movie theaters, and airports. Complete with an internal computer, sound card, and video graphics monitor, these dazzle stations can be a powerful advertising tool — especially when they have motion detection circuitry that triggers a video presentation at the precise moment an unsuspecting patron comes near.

This article explains how to construct a USB motion detector that will give your computer this hey-you ability, using a free Windows presentation applet I wrote, USB Multimedia Presenter, so that you can start your own kiosk advertising campaign. You can also use the setup for practical jokes, or just to amaze or amuse your friends.

To interface between the detector and computer, I used an off-the-shelf USB device which requires no drivers to install, since it uses existing Windows drivers. The detector draws all the power it needs from the computer, which further simplifies things. All the parts for the project are easy to find, and if you have basic soldering and mechanical skills, you can put it together in a single evening.

Connect the Detector to the USB Interface

Drill a ³⁄₁₆" hole in the top of the motion detector enclosure, to accommodate the USB cable. Center

DIY HOME

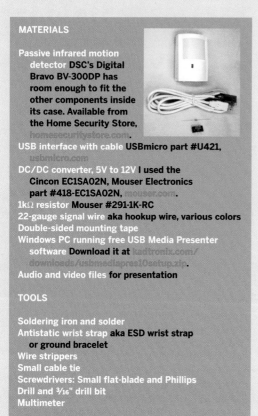

MATERIALS

Passive infrared motion detector DSC's Digital Bravo BV-300DP has room enough to fit the other components inside its case. Available from the Home Security Store, homesecuritystore.com.

USB interface with cable USBmicro part #U421, usbmicro.com

DC/DC converter, 5V to 12V I used the Cincon EC1SA02N, Mouser Electronics part #418-EC1SA02N, mouser.com.

1kΩ resistor Mouser #291-1K-RC

22-gauge signal wire aka hookup wire, various colors

Double-sided mounting tape

Windows PC running free USB Media Presenter software Download it at kadtronix.com/downloads/usbmediapres10setup.zip.

Audio and video files for presentation

TOOLS

Soldering iron and solder

Antistatic wrist strap aka ESD wrist strap or ground bracelet

Wire strippers

Small cable tie

Screwdrivers: Small flat-blade and Phillips

Drill and ³⁄₁₆" drill bit

Multimeter

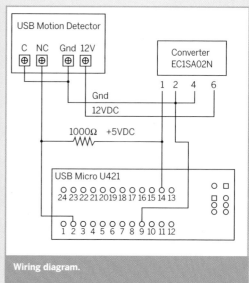

Wiring diagram.

the hole on the seam where the 2 halves of the enclosure meet. Do not let the drill bit penetrate more than ½" inside the unit, as this can damage internal components (Figure A).

To open up the detector, first unscrew the retaining screw, if the unit has one. Then firmly grasp the unit in your left hand and use your right hand to pull the halves apart while gently pushing down on the retaining tab with a small flat-blade screwdriver. Set both halves aside.

At this point you'll start handling static-sensitive electronic components, so you are strongly advised to wear an antistatic wrist strap clipped to a suitable electrical ground. This will protect the components from damage from electrostatic discharge.

Use a Phillips screwdriver to remove the screw that holds the detector's printed circuit board. Carefully remove the PCB and set it aside.

Solder short (2") lengths of signal wire following the project wiring diagram (above) to pre-wire the USB interface, DC voltage converter, and 1kΩ

resistor. Note that the 4 pins of the SIP package version of the DC converter are numbered 1, 2, 4, and 6, to match the pin numbering on other versions of the component. Leads run from terminals 9 and 14 of the USB interface PCB to pins 2 and 1 of the converter, respectively. Pin 1 of the converter connects to the resistor, and pins 2 and 4 connect together. Solder leads to converter pins 2 and 6, and connect another lead from USB PCB terminal 2 to the unconnected end of the resistor, but leave them loose for now (Figure B); we'll attach them when reinstalling the detector PCB.

Arrange the pre-wired components inside the detector case, and secure them to the back shell with double-stick mounting tape. Be careful not to obstruct the area where the detector PCB will be reinserted. The components should fit just under the detector PCB without making contact (Figure C).

Now, reinstall the detector PCB and secure it with the small original screw. Complete the wiring as shown in the wiring diagram: USB PCB terminal 2 and resistor to terminal NC, DC converter pins 6 and 4 to 12V and Gnd, and terminal C bridged to Gnd (Figures D and E). The PCB has screw terminal blocks, so you don't need to solder. The NC (normally closed) and C (common) terminals serve as a switch that turns things on and off, and the USB interface routes this signal to the computer.

Close the detector case back up; that's it for the hardware!

Photograph by Ken Delahoussaye

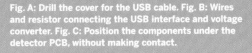

Fig. A: Drill the cover for the USB cable. Fig. B: Wires and resistor connecting the USB interface and voltage converter. Fig. C: Position the components under the detector PCB, without making contact.

Fig. D: The wire connections to motion detector PCB terminal blocks. Fig. E: The finished circuit with the motion detector PCB back in place.

Photography by Ed Troxell (Figures A–B) and Ken Delahoussaye (Figures C–E)

Configure the Software, and Run with It

The software, USB Multimedia Presenter, builds a playlist with all the audio and video files in a specified folder. Then, each time someone comes near the detector, it randomly selects and plays one of them. To get started, download the application from kadtronix.com/downloads/usbmediapres10setup.zip onto your PC, then unzip it, install, and launch.

In the Configure box, choose a folder on your system containing the media files you want to trigger; this defaults to the *C:\WINDOWS\MEDIA* folder, which is ideal for testing purposes if you don't have a specific playlist already.

Test the software by clicking Start, with the Use Trigger Device box unchecked below. This will shuffle-play the files in "demo mode," with a set delay between each.

Now let's put the system together. Plug the motion detector into the computer and position it in a direction that will detect movement. To eliminate false-positive triggerings, you might strategically limit its field of view. My detector picked up on adults moving up to 15 feet away.

In USB Multimedia Presenter, check Use Trigger Device at the bottom of the pane, then click Start.

You're up and running. The Device Activity Indicator should flash from green to red when motion is detected. Approaching visitors will be amazed as your presentation magically begins for them.

Create a Presentation

For the presentation itself, you can do anything you want. Brightly colored graphics, flashy animations, and special effects are always effective eye-catchers, and once you've got an audience, the content must hold their interest long enough to convey the message.

There are many other possible uses for the system. Once, I set the detector up just outside my front door, so that the computer would alert me when a visitor approached. Another time, I set it up to play a *.wav* audio of a shattering window. For added effect, I turned up the volume on my computer speakers. Whenever the trigger occurred, it had everyone in the house running around looking for broken glass.

Ken Delahoussaye (kdelahou@kadtronix.com) is a software consultant in Melbourne, Fla., who specializes in embedded software and PC applications. He operates kadtronix.com, which features access control and related resources.

RELATIVE MEASUREMENTS

Making good guesses using what's at hand (including your hand). By Donald Haas

Having the right tool for the job is important, but sometimes you have to work with what's available. Relative measurements can be just as accurate as standard measurements. Memorize the dimensions of a few common items that you usually have with you or readily available, and you're set.

I know that with my arms stretched out to the sides, my fingertips are 6' apart; great for measuring rope or a rough estimate of a room size. A dollar bill is 6⅛"×2⅛", a credit/debit/gift card is 3.370"×2⅛", a business card is 3½"×2", a quarter has a 0.955" diameter. Most floor tiles are 12"×12" while ceiling tiles are 2'×4'. Remember, it's all relative.

Measure This: Far Yard Sale

The Problem: The Mrs. calls from a yard sale where she found a desk she thinks would be perfect in a nook in the office, but she isn't sure it will fit. She wants me to measure the nook, then meet her at the yard sale and measure the desk. The yard sale is over a half-hour drive away. She doesn't have a tape measure, and neither does the owner of the house.

The Solution: I have her measure how many credit cards (using the long side) wide and deep the desk is, while I measure the nook with the length of a credit card. The nook is 3'0" or 10.7 credit cards wide, while the desk is 11.25 credit cards wide, or about 3'2". Too big. I save myself an hour on the road and get back to playing with my soldering iron.

Measure This: Cookie Calculation

The Problem: I receive a wonderful old Swedish recipe for gingerbread cookies, and it calls for 4 hectograms of flour. I have to check my copy of the *MAKE Pocket Ref*, which gives me the conversion factor of 0.22 (4 hectograms = 0.88 lbs.). That's when I remember I don't own a kitchen scale.

The Solution: I remember the phrase "A pint's a pound the world around," which refers to a pint of water (16 fluid ounces, actual weight is 1.04 lbs). I have a measuring cup that's marked in pints, so I measure out approximately 0.88 pints of water. I place that in 1 of 2 identical large cups. Then I spoon flour into the second cup until their weights feel equal to me. Approximations like this are fine for the cookie recipe — they're delicious.

Donald Haas (maker@kurtroedeger.com) is an amateur tinkerer and perpetual daydreamer. He's always looking for new solutions to old problems.

Make: TIPS!

Lubricating Saw Blades
Before cutting fret slots, or much of anything else, rub an old candle along the edge of your saw blade. You'll be surprised how much easier it makes the process to have a bit of lubrication.
—*Frank Ford*, frets.com

Find more tools-n-tips at makezine.com/tnt.

Illustration by Alison Kendall

PAULOWNIA ARCHERY BOWS

 Making stuff with the wood that just might save the world. By Dan Albert

Photography by Dan Albert

When we took possession of our humble London home, I was shocked to find that all the window treatments had been removed. So we suffered the rat-in-a-maze Ikea gantlet to get a good price on new Venetian blinds. I hung the new blinds immediately but it took me months to get around to tailoring them by removing the extra slats. As soon as I did, I realized that I had a maker's trifecta win in my hands: easily worked hardwood, prefinished and free.

First I built a new box for our kitchen plastic wrap, then my daughter wanted some doll furniture. Next was a laminated beam to repair our baby stroller, and a few slats to serve as drawer dividers for the clothes dresser I'd built ages ago but never quite finished. But the *pièce de résistance* was a set of archery bows that I whipped up to the delight of the neighborhood kids.

Wonder Wood

It turns out that Ikea's Lindmon blinds (product #10092570 at ikea.com/us) are made from *Paulownia elongata*, an incredibly fast-growing hardwood that originates in southern China and Southeast Asia but has been bred for cultivation around the world, including in the somewhat colder climate of the southeastern United States (see paulowniatrees.org).

Paulownia is made into everything from coffins to stringed instruments. The wood is fine-grained, virtually knot-free, and easily worked. It has the look and feel of balsa, only slightly heavier with about twice the strength and hardness.

I could tell the wood was something special, but claims that paulownia is an environmentally friendly "wonder tree" made me skeptical, so I did some homework. I checked out the scientific literature

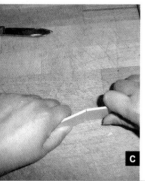

Figs. A and B: Paulownia artifacts from the World Paulownia Museum in Tokyo, Japan. Fig. C: If you don't have a fine-tooth saw, it's easy to score and snap slats of paulownia wood.

Figs. D and E: A plastic wrap dispenser made from scrap 1"×2" lumber, a dowel, and a few paulownia slats. Remove the serrated edge from the plastic wrap's packaging and attach it to your box.

and even paid a visit to Tokyo, where the Tanaka furniture company maintains the World Paulownia Museum adjacent to its production facilities (Figures A and B) — visit them online at kiriyatanaka.co.jp (and for an always amusing English translation use babelfish.altavista.com).

After researching, playing a paulownia guitar, and seeing evidence of the wood's fire resistance, I'm a believer.

The tree is sustainable, reduces soil erosion, and grows up to a very marketable 12 meters high in just seven years. After harvest, new trees grow from the stumps. Planted alongside food crops, it boosts yields by creating a windbreak and an improved microclimate. It serves as a biomass fuel and has been shown to grow nicely in swine lagoons, where it provides waste remediation. It has a very high ignition point and is relatively rot-resistant. And apparently it was named for a Russian princess.

Working with Paulownia

Dimensional paulownia lumber is slowly becoming available in the United States, but Ikea has helpfully processed a bunch of it into 1⅜"×⁵⁄₆₄" (35mm×2mm) slats complete with precisely located holes good for making all kinds of stuff. There are plenty of extra slats

after shortening the blinds, but it might even be worth buying blinds just for the slats: they run 6.25 cents per linear foot, and you get the mechanism as a bonus.

Here's a look at some of the techniques of working paulownia slats. I mostly use a dovetail saw, simply a small back saw with fine teeth. You can find one at the hardware store for $10–$15. Any saw will do, but the fine teeth make clean cuts and let you easily hold the piece steady with your hand.

No saw? You can score and snap (Figure C) or even cut all the way through the slats with a pocket-knife or utility knife.

Gluing can present a problem, as there is a thin layer of lacquer on the slats. For a strong joint using carpenter's glue, sand the slat to expose the bare wood. The coating seems light enough that polyurethane glue worked well even without sanding (I like Gorilla Glue because it's available in small quantities).

Quick Paulownia Projects

My first project was a quick-and-dirty plastic wrap dispenser made from scrap 1"×2" lumber, a dowel, and a few slats (Figures D and E).

Next, my daughter and I built some doll beds. We used a bit of tree branch to give a rustic look, and bent the slats to create a nice Gothic arch (Figure F).

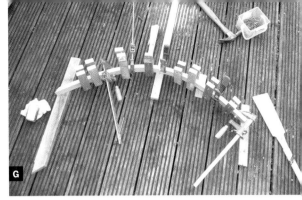

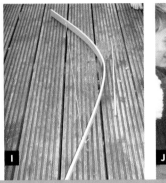

Fig. F: A doll's bed for my daughter, Molly, age 7. We used a bit of tree branch to give a rustic look. Pre-drill the slats to avoid split ends. Fig. G: To make a laminated beam, spread a thin layer of glue on each slat and clamp together. Fig. H: The finished beam can easily support 60 lbs. Fig. I: String 3 slats with a length of line to make a bow. Fig. J: My son, Joey, age 4, shows that bamboo sticks make excellent arrows, even without feathers.

The kids had broken the footrest on our baby buggy, but I could never find anything both rigid and thin enough to effect a repair. Then I hit on the idea of a curved, laminated beam of paulownia slats. My 8"-long, 3-slat-thick beam with a slight curve has held up better than the original plastic.

The strength of this little beam was so impressive I decided to experiment with something bigger. I hammered nails into our deck where the center and 2 ends of the curve would be. Dry-fitting the slats proved this simple jig would work.

I then wet down 1 side of the slats and spread a thin layer of Gorilla Glue on each before setting them into my jig. I clamped the whole thing together with all the clamps I had, plus temporary clamps made from lumber and drywall screws (Figure G). It's important to apply pressure all the way along your glue-up.

Once dry, the beam measured 37" long by 1" thick and had a chord (height) of 7". It supports at least 60 lbs. with little deflection (Figure H).

Archery Bows

My beam experiment and quick projects were fun, but the best use of paulownia is also the simplest: archery bows. All you need are 3 slats and a bit of low-stretch cord — nylon clothesline works great.

Cut a length of line about 1' longer than your slats. Tie a stopper knot at one end, then stack your slats and thread the line through 1 set of holes. Bend into a curve, thread through the other set of holes, tie off, and you're done (Figure I). Add duct tape to show the kids where to hold the bow (Figure J). The quick construction is key, because as soon as the first kid gets one, the others will be all over you.

For arrows, you can use any straight sticks; bamboo garden stakes work well. Cut a vee groove at one end to accept the bowstring. You would think that you'd need feathers to get the arrows to shoot well, but in fact the sticks fly nicely naked.

If they flew any straighter they'd be deadly, which isn't really the point. Similarly, you could add more slats to your bow for more force, but try not to start an arms race.

Dan Albert (exchaoordo@gmail.com) is an assistant professor of history at Salem State College in Massachusetts. He writes on transport technology.

The Spinning Cylinder Illusion

This is a toy, a puzzle, and an illusion all in one, with only one moving part. No batteries are required. I like simplicity, especially when it produces puzzling complexity.

» I've not been able to track down the origin of this homemade toy, but it isn't very well known outside the community of physics teachers. It's a kinetic illusion, one that depends on physical motion to make you see something that isn't there.

A familiar example of a kinetic illusion is the strobe effect sometimes seen in old movies, causing the spokes of a carriage wheel to seem to be turning in the wrong direction.

In its simplest form, this toy consists of a hollow cylinder of rigid plastic. The version in these photos is 4cm long and 1cm in diameter, with 2mm wall thickness. It was cut from a piece of polyethylene plastic tubing that happened to be lying on my workbench. Whatever tubing you use, be sure to choose a very straight piece.

This material can be cut easily with a single-edged razor blade, scrap of board, and hammer. Use heavy gloves and goggles in case the razor blade breaks. Or use a hacksaw, then smooth up the cut with fine sandpaper. Cut the piece initially a bit long, then shorten it to optimum length by trial. I've made small ones in 2-, 3-, 4-, and 5cm lengths. I've also made them from the barrels of old ballpoint pens, and larger ones from PVC plumbing pipe.

Glue or paint a red dot near one end and a green dot at the other end, or use a felt-tip marker to make distinctive symbols at the ends.

The Vanishing Dot

Place the cylinder on a flat surface. It's best to choose a surface that won't dent or scratch, like a smooth, hard floor. Press down on 1 side of 1 end with your thumb, as shown in Figure B, until the cylinder slips away from your thumb and flies off spinning. It will settle down, revolving about its midpoint and at the same time spinning about its long axis (Figure A).

Here's a sequence of photos showing the cylinder beginning to spin (Figure C), not quite settled down (Figure D), and finally spinning in one place with nearly constant angular speed (Figure E). These photos were taken with a digital camera, but they show approximately what appears to the eye.

All these photos are taken from the same camera position and distance, and reproduced at constant scale. Therefore lengths on the pictures may be directly compared. Notice that a stable illusion showing 4 equally spaced dots occurs when the cylinder rotates about its center, and both ends traverse the same circle, which has a diameter equal to the length of the cylinder.

The puzzling outcome is that when the cylinder settles down to a uniform spin rate at a particular location on the table, you see only one color dot (red, in this case), repeated N times, where N is the ratio of the cylinder's length to its diameter.

The dot you see is the one at the end you pressed with your thumb to launch it. You don't see any evidence of the dot at the other end (green, in this case). Launch the cylinder by pressing the green dot end and then you see only the green dot, and not the red one.

How can we explain this unexpected behavior? Note that translational motion of the cylinder's center of mass stops just before stable dot patterns become visible. Note also that what we call the "stable motion" occurs when the ends of the cylinder trace a circle.

When you think you've got it figured out, test your understanding by considering this follow-up question. If you spin this toy on a glass sheet, and someone looks up at it from below the glass, which colored dot would that observer see? A glass-topped coffee table is good for this experiment.

What would be the optimum length-to-diameter

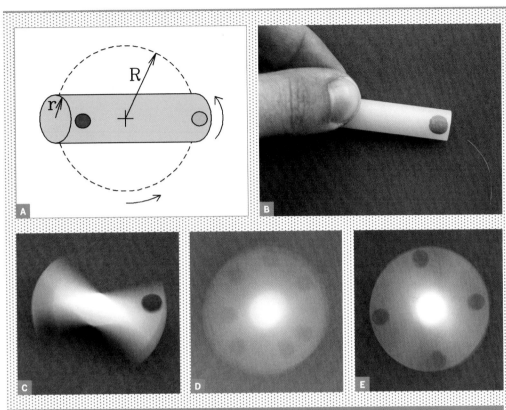

Photography by Sam Murphy; illustrations by Donald Simanek

WHERE DID THE GREEN DOT GO? Fig. A: The spinning cylinder revolves around its midpoint (radius R) and simultaneously around its long axis (radius r). Fig. B: A small plastic cylinder has a red dot on one end, and a green dot on the other. Fig. C: Immediately after spinning, the motion is chaotic. Fig. D: As the cylinder stabilizes, many red dots are visible. Fig. E: When the spinning has stabilized, 4 red dots are clearly visible. The green dot is not visible.

ratio for a given number of dot repetitions, if you used a *solid* cylinder? (It may not be the same.) What role does moment of inertia play here? Friction? Energy? There's a lot of physics here to address, and it may not be easy to do.

Some Curious Observations

A 3:1 version with ¾" PVC pipe works best for the following experiments. This larger version also seems to exhibit less translational motion from the point of launch before stabilizing, enabling you to do the experiments on a smaller surface area.

Spin the 3:1 cylinder by launching it forcefully at the red dot end, and observe it from the side. For a while, it spins with the red dot end rather high in the air. During that time you may see a brief but stable pattern of 5 red dots. The cylinder's spin slows until

its red dot end is only slightly above the table. Then you see a pattern of 3 red dots for a longer time. During the transition you do *not* observe a stable pattern of 4 dots, except perhaps fleetingly. Why 5 and not 4? Could it be that the lower end is still sliding on the floor? This would be consistent with the observed shorter time that this motion is stable.

Solution

Another thing I like about this toy is that it combines physics and physiology. Initially, because of the manner of launching, the cylinder spins about its own long axis and slides on the floor (Figure F, next page). Because the end you launched (red dot) is also propelled sideways, the other end (green dot) drags on the floor. The friction with the floor slows the cylinder's motion and also changes its direction.

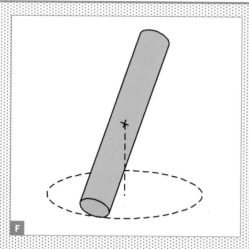

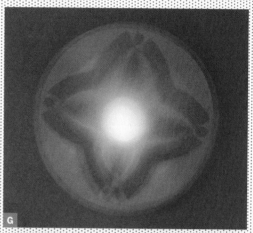

Fig. F: As the cylinder spins about its long axis, it traces a circle around the floor. When the green dot is up, the dot is moving too rapidly to be distinctly seen. The red dot moves most slowly when it is facing upward, and is therefore briefly visible, as seen from above. Fig. G: Different decorations can produce kaleidoscopic patterns. Fig. H: Experiment with patterns using colored tape.

The motion stabilizes with both ends moving around a common circle, but the end you launched is not quite touching the floor, while the other end is rolling around the circle without slipping. When the green dot is up, the dot is moving too rapidly to be distinctly seen. When the red dot is up, it's moving around the cylinder's long axis in the opposite direction to the direction that end of the cylinder moves around the larger circle. Therefore the red dot moves most slowly when it's facing upward, and is therefore briefly visible, as seen from above.

If the cylinder is 4 times as long as its diameter, the larger circle is 4 times the diameter of the cylinder. When rolling without slipping is established, the cylinder rotates about its long axis 4 times for every time it spins around the large circle, and therefore we see the red dot 4 times.

When the cylinder is spun very forcefully, it may for a short while spin with the cylinder elevated at a large angle, but it soon settles down to the motion described. There's a lot of good physics going on here, involving moments of inertia, gyroscopic motion, rotational energy, and so forth. If anyone wants to work that out, please get in touch with me at dsimanek@lhup.edu, and I might find space for your analysis on my web page.

More Illusory Fun

Now that you understand the principle, you can get creatively arty. If you use a larger-sized white plastic tube, you can decorate it in various ways so that when it's spun, it will produce "kaleidoscopic" patterns (Figure G). Star patterns are easy. Colored flexible plastic tape is good for initial experimentation (Figure H), and more permanent marking may be used later.

Donald Simanek is emeritus professor of physics at Lock Haven University of Pennsylvania. He writes about science, pseudoscience, and humor at www.lhup.edu/~dsimanek.

Photography and illustration by Donald Simanek

The Electronic Frontier Foundation (EFF) is the leading organization defending civil liberties in the digital world. We defend free speech on the Internet, fight illegal surveillance, promote the rights of innovators to develop new digital technologies, and work to ensure that the rights and freedoms we enjoy are enhanced — rather than eroded — as our use of technology grows.

PRIVACY EFF has filed two major lawsuits — one against the government and one against AT&T — challenging the NSA's illegal warrantless wiretapping program. *eff.org/nsa*

FREE SPEECH EFF's Coders' Rights Project is defending the rights of programmers and security researchers to publish their findings without fear of legal challenges. *eff.org/freespeech*

INNOVATION EFF's Patent Busting Project challenges overbroad patents that threaten technological innovation. *eff.org/patent*

FAIR USE EFF is fighting prohibitive standards that would take away your right to receive and use over-the-air television broadcasts any way you choose. *eff.org/IP/fairuse*

TRANSPARENCY EFF has developed the Switzerland Network Testing Tool to give individuals the tools to test for covert traffic filtering. *eff.org/transparency*

INTERNATIONAL EFF is working to ensure that international treaties do not restrict our free speech, privacy or digital consumer rights. *eff.org/global*

EFF.ORG

ELECTRONIC FRONTIER FOUNDATION

Protecting Rights and Promoting Freedom on the Electronic Frontier

EFF is a member-supported organization. Join Now! *www.eff.org/support*

Magnetic tacks, a doorbell for your cubicle, science for the hungry, and lessons from the Game of Life.

TOOLBOX

Hooked
Micro and Pocket Grappling Hooks
$22-$27 countycomm.com

You will never, ever need your own personal, pocket-sized grappling hook. Ever. Until that one crowning moment when disaster strikes; someone will yell, "If only we had a grappling hook!" You'll coolly draw your Micro Grappling Hook from your pocket, unscrew its base to reveal its three tiny spikes, thread them into their angled holes, recap it, and without saying a word, swing the hook toward its target, saving the day. I know all this because I lived it — at my son's fifth birthday party right after I tossed his new airplane onto the roof.

Designed for the more serious needs of military personnel, the microhook is beautifully machined from 300 series stainless steel. The end hole will accommodate 550 paracord, but for long-range grappling, soldiers use fishing line and a Pocket Fisherman. Your own hero moment is waiting for you.

—*John Edgar Park*

Want more?
» Check out our searchable online database of tips and tools at makezine.com/tnt.
Have a tool worth keeping in your toolbox? Let us know at toolbox@makezine.com.

MagTacks

$7 makezine.com/go/magtacks

Ever wanted to tack something up but not mar it with permanent holes? MagTacks are the answer to your prayers. Each one is a magnet and a tack rolled up in one compact, nonslip package. Simply push the tack end into your corkboard or wall, then separate the two magnetic sides and use them to hold your item in place. Perfect for displaying (while preserving) photos, schematics, plans, posters, you name it. You can even get crazy and reverse the usage to "tack" something to your fridge. And when you're using only the magnet, you can flip the tack around to store it safely in its own cushion. I heart simple ingenuity! —Goli Mohammadi

MoteDaemon

Free screenfashion.org

Screenfashion has released MoteDaemon, an application that allows Adobe Flash/Flex developers to create Wiimote-controlled applications for OS X. Imagine the potential for homebrew video games and VJ applications.

Pairing the Wiimote to MoteDaemon is a snap. (If you aren't comfortable with ActionScripting in Flash, you won't be able to get much immediate use out of this software.) Fortunately, a little test-drive application, WiiCockpit, is included, and although the documentation is in German, they were kind enough to include the .fla document with all the ActionScript for WiiCockpit. Experienced Flash developers will quickly figure out how to integrate MoteDaemon from a peek at WiiCockpit's code. **—Bill Byrne**

Mini Metal Lathe

$525 grizzly.com

I don't have a lot of room or money to spend on fancy machine tools, but I really enjoy making things out of metal. This meant I needed a lathe. Not a wood lathe or a pen lathe — I *needed* a metalworking lathe.

Lucky for me, I found an entire community centered on 7"×10" and 7"×12" mini lathes. You can buy them from many sources; mine is from Grizzly Industrial, Inc. It's small, but it can do quite a bit. Plus, there are many people who are constantly upgrading or hot rodding their machines to get the most out of them. I used mine in the first month to repair a pulley on my ancient garage door for which there were no replacements. I figure I saved the cost of the lathe right there. And I plan on using this for years to come to make all sorts of things. —Brian Graham

Life Lessons from the Game of Life
Conway's Game of Life Kit
$18 adafruit.com

Our family reunions have become quite an affair now that my mother can tally 21 grand-kids. Aged from 1 to 20, keeping them busy is a challenge. For a previous reunion, I had built a soda bottle rocket launcher that was a big hit (see MAKE, Volume 05, page 78). This year I had a small LED craft project for the little kids, but needed something more challenging for the older kids. Luckily, I came across the Game of Life kit that Lady Ada at Adafruit Industries had just updated.

The Game of Life, as proposed by John Conway in 1970, is not really a game — other than it's fun to watch. From a given initial pattern on a grid, using simple rules governing the life or death of each square, amazing shapes known as cellular automata can evolve. The larger the grid, the more elaborate the patterns can become. Performed with paper and pencil in Conway's time, the game is a perfect application for a small microcontroller driving an LED array. I purchased two of the kits and had my girls, ages 9 and 13, try to put them together. They both picked up soldering quickly, and in a matter of hours we had two working boards. What's more, they enjoyed it!

At the reunion, we had a table set up with two soldering stations. I pulled kids aside two at a time, gave them a brief lesson on solder-ing, and then let them try their hand at it. I had purchased a selection of LEDs so they could choose their own colors. Some of them picked up soldering quickly, and rapidly finished their kit. Others took their time, and would take breaks as their attention waned. Some required several sessions to finish the kit. Inevitably, someone would make a solder bridge or put an LED in backwards, and a lesson on using desoldering braid was taught.

As each kit was finished and tested, it was attached to the other boards that were already assembled. The boards communicate their status to each other, allowing larger and more elaborate patterns to be formed. At the end, we had ten boards, or 160 LEDs, working together.

We put the entire assembly in a prominent location and let it run for the remainder of the reunion. It was as mesmerizing to watch as the campfire was. Several times, large repeating patterns evolved.

As our families prepared to go their separate ways at the end of the reunion, the kids were given their boards to take home. Hopefully one of the lessons learned was how reunions are a chance to come together, if only for a while, and be a part of something larger than ourselves.

—Ken Olsen

Watch the video of our ten Game of Life boards at makezine.com/go/reunion. The younger kids' project, Buggy, can be seen at makezine.com/go/buggy.

ThingamaKit

$55–$65 bleeplabs.com

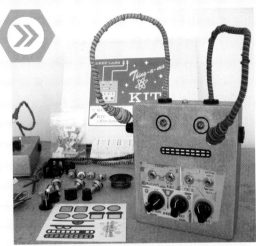

With his deranged-but-cute robot good looks, the ThingamaKit analog synthesizer is great fun to build, and even more fun to play with. Whether I'm trying to make it sound like a lounge act from the future or a game of *Space Invaders*, I can't seem to put it down.

The kit is beautifully designed: a clearly labeled circuit board, bagged and tagged components, and excellent printed instructions, complete with full-color photos of the build process. You supply basic soldering skills, a few tools, and an evening or two.

ThingamaKit has knobs to tune the waveform rate and shape of his main oscillator. This by itself sounds roughly like emergency-room equipment on the fritz.

His true power is revealed when you look into his light-sensitive "eyes," a pair of photoresistors. The amount of light he sees with his right eye adjusts pitch, which means you can wave your hand around to play him like a theremin. For bleepier sounds, use his blinking "LEDacle" appendage as a light source. His left eye and its corresponding LEDacle, switches, and knob adjust the modulator, which affects the tempo of the bleeps and adds audible sidebands at higher frequencies.

The net result is a space-age sonic freak show. ThingamaKit is both a satisfying build and an adorable, noisy companion. —*John Edgar Park*

Wave Shield for Arduino Kit

$22 adafruit.com

I'm fascinated by electronics and I'm always looking for practical ways to learn about them. It's not enough to read about how things work; sometimes I just need to get my hands dirty. Lately I've been hacking on an Arduino for a rudimentary cocktail-serving robot, and I decided I needed to get some sound on board. Some research brought me to the Wave Shield kit sold by Adafruit Industries.

The kit is a PCB "shield" that piggybacks onto your Arduino to add audio playback. It plays 22kHz, 16-bit *.wav* files from an SD card through a standard headphone jack.

The Wave Shield is a great, low-barrier-to-entry way to get sound into your project. Unfortunately, the Wave Shield does take up a sizable chunk of RAM to buffer the audio on the Arduino, but overall it's a neat little kit to work with.

Assembly and soldering are a snap; it took about 20 minutes to get it ready to plug in and test. On the programming side, Adafruit has a fairly extensive library to download, as well as a bunch of code examples on the site.

As long as you aren't expecting hi-fi audio quality I think you'll be pleasantly surprised by the Wave Shield kit, and at 22 bucks it's worth it just to have your Arduino compliment you on your soldering skills.

—*Mitchell Heinrich*

« Desktop Earbud Speakers perpetualkid.com

A real attention-getter for your desk, these 500 XL Desktop Earbud Speakers look exactly like a pair of iPod earbuds — except they're 500 times bigger! The sound is not knockout great but it's impressive for their size and price, and people will freak out when they see them on your desk. They've got a built-in amp, and can be powered by batteries, USB, or a standard wall-plug power supply. Podcast: twit.tv/dgw528

« Mini Surge Protector Charger belkin.com

So you have just two outlets next to your desk, and way more than two devices to plug in. This surge protector is the perfect answer. Plug it into one outlet and you get three outlets, plus two USB charging ports! It has a rotating plug so you can face the device in the direction that's most convenient. A green LED lets you know the outlet is powered and the surge protection is on. Podcast: twit.tv/dgw534

« Cubicaller cubicaller.com

When someone comes to your cubicle and you're involved in your work, how do you know they're there? I installed the Cubicaller. It comes with self-adhesive velcro tape so it installs in a minute. You can choose from 12 different sounds such as doorbell, duck quack, fanfare, and meow. Podcast: twit.tv/dgw381

« Socket Sense socketsense.com

Hate those big "outlet killer" power adapters that take up the space of two outlets? Here's a surge strip that expands! You can extend the strip sockets by about 8" by pulling on both ends. The adjustable sockets are also set on an angle to make sure a transformer plug does not block another socket.

« OpenIt! enjoyzibra.com/openit

If you hate those almost-indestructible plastic blister packs that so many new small electronics are packaged in, check out the OpenIt! It looks like pruning shears but instead of small branches, it cuts through heavy-duty plastic shell packages. There's also a retractable box-cutter built into one side of the handle, and a mini screwdriver in the other, so you can open the "kid-proof" battery compartment of your new gadget. Podcast: twit.tv/dgw496

« Stealth Switch thinkgeek.com

Playing games at school? Don't want to get caught? Stealth Switch is the answer. Install the software and plug the hidden foot switch into an available USB port, then just a tap of your toes will hide the current window, hide all windows, or hide all except the one you want! (For PC only.)

Dick DeBartolo wears two hats, which is why he looks so strange when you meet him. He's had something published in every issue of *MAD* for the past 40-plus years! His other career is as the Giz Wiz, as in Gizmo Wizard. Dick does the Daily Giz Wiz podcast, along with tech genius Leo Laporte. Check out Dick's gadget reviews at gizwiz.biz and follow him at twitter.com/thegizwiz.

« Play with Your Food

The Hungry Scientist Handbook by Patrick Buckley and Lily Binns
$17 Collins Living

Ever wonder what the original do-it-yourself handbook was about? I think it must have been about food. Accordingly, in their new book, Patrick Buckley and Lily Binns celebrate hackers in the kitchen.

This book is about playing with your food. It's got recipes for kitchen- and party-related hardware as well as edibles, and includes juicy snippets of technical information related to the projects. This is a bit like how your favorite science textbooks had those little inset "relevant project" sections that were the best part — except in reverse, with adventurous projects as the main topics, and tidbits on the science behind it all in the insets.

The projects can do delightful things that food shouldn't do (like blink), and they cover a wide range of complexity and required tinkering time. The crafty among you will be tempted to try every project at least once, and some projects may even find their way into your kitchen ritual.   —Kenny Cheung

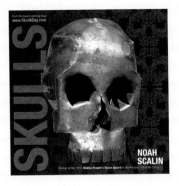

« Kingdom of the Skull

Skulls by Noah Scalin
$15 Lark Books

The genius and arty delight of the Skull-A-Day website can now be had in a new book! Published by Lark Books, *Skulls* is sure to be a coffee table favorite all year long. The book showcases some of Noah Scalin's creations during his yearlong adventure in which he made a new skull each day and posted it on his award-winning blog, skulladay.com. Scalin's experimentation knew no bounds; he made skulls from keyboard keys, pennies, rice, a watermelon, a plastic bag, a light bulb, pancakes, wire, even a single acorn (one of my favorites). The book captures the fun and brilliance of the website — turning the pages is an exercise in oohing and ahhing at one outrageous skull after another.   —Shawn Connally

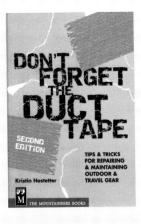

« Stuck on Camping

Don't Forget the Duct Tape by Kristin Hostetter
$8 Mountaineers Books

This little book should be standard issue with every new piece of camping equipment sold. It's a great collection of tips for keeping outdoor and travel gear in tip-top shape, full of smart advice for repairing broken zippers, snagged sleeping bags, and sputtering camp stoves. Even more importantly, it lays out the best way to keep problems from happening in the first place. Kristin Hostetter is the gear editor at *Backpacker* magazine, and you can tell she's seen a lot of gear. Heck, it's a great thing to take with you when shopping for your stuff in the first place.   —Arwen O'Reilly Griffith

Gamer's Dream
$10 garrysmod.com

As a tinkerer and a gaming fanatic I was pleasantly surprised to find my two passions merged into one glorious, 3D-rendered package. *Garry's Mod* is a PC gamer's/maker's dream, a sandbox environment that incorporates the interactivity of *Half-Life 2* and pops the lid off its can of confinement.

Tools like Easy Weld let you piece together the most obscene creations. Ever wanted to make a mech out of filing cabinets, paint cans, and coffee tables, or see what a rocket-powered hovering toilet with mounted machine guns looks like? Of course you have.

GMod also lets you adjust gravity, light, color, proportions, and movement to basically shape your own world. Physics does come into play, which adds a humorous reality to some of the misshapen monstrosities you can throw together. You're free to make mistakes, trash your prototype, even obliterate your avatar, and all you have to do is make a few clicks to right the world again. Riders of the God complex highway, here's your stallion.

—*Ryan Beacom*

Get the Hook
The Amazing Monkey Hook
$3 for 2 monkeyhook.com

My wife is an artist, so I've hung many pictures, and I thought I'd seen all the available systems. On a recent trip from Australia to visit our son and his wife, I was given the job of hanging two paintings brought by my wife on the flimsy sheetrock walls of their home.

Whilst I contemplated the many wall fixings at Home Depot, a young man, sensing my dilemma, handed me a pack of Amazing Monkey Hooks, which he had used before. I read the description on the pack and realized that here was a device that in engineering terms was perfect.

It would spread the hanging load over the maximum area of the fragile wall (in compression — its strongest load-bearing orientation). It would make a minimal hole in the wall (no drilling — the hook penetrates with a little manual help). It would maintain its correct orientation (thanks to its bent profile that presents the pointed tip to the backside of the wall sheet). And it could be easily removed, leaving only a small hole for repair.

I came 8,000 miles to find it, and I took home a future supply! —*Ross Griffith*

Tricks of the Trade By Tim Lillis

Making labels is clearly easy.

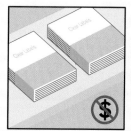

Need to print some clear labels but put off by the high price tag of commercial label paper? Try this trick from Jeff Crews for a DIY solution.

First, take a used paper label sheet and remove all the borders. You can save these sheets in a file after you use them, so you have them at the ready.

Next, insert the sheet into your laser printer so the glossy smooth side gets the toner. Print!

Take the printed sheet, and mount tape or clear contact paper on top of the printed area. Carefully peel it back, and the toner will adhere to the clear material. Attach and enjoy!

Have a trick of the trade? Send it to tricks@makezine.com.

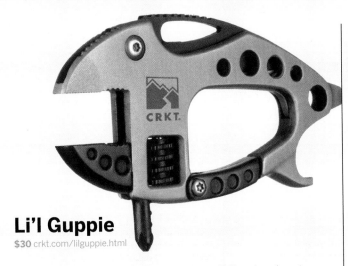

Li'l Guppie

$30 crkt.com/lilguppie.html

Why the heck do I need another pocket tool? I've already got a multi-tool with pliers, knives, screwdrivers, and more. Well, the reason is this: ever try to loosen a bolt without an adjustable wrench? It's pretty hopeless; the nonparallel jaws of pliers just don't cut it. The Li'l Guppie is the answer; it has an adjustable wrench on it.

Thanks to its carabiner-like form, I can hang it from my belt loop, ready to be deployed on a stubborn bolt. It's got a fangy little blade that makes me nervous (I wish it locked in place), but turning the thumbscrew to open the wrench reveals a Phillips screwdriver, an extra clever feature. A flathead screwdriver is cast into the butt of the tool, with a pretty serviceable bottle opener built in, too.

—John Edgar Park

HobbyCNC Pro Driver Board

$80 hobbycnc.com

I'm a mechanical type of guy. So when I decided to refurbish a desktop CNC milling machine, I was a little nervous about the electronics, especially since the existing electronics were falling apart. I could have tried to build a stepper motor controller, but I didn't want to spend all that time and money on a refurbish, only to rely on my weakest skill at the very end to see if it worked.

Then I found HobbyCNC. You can get different levels and kits to fit your needs. They're easy to assemble for anyone with basic soldering skills, and there's plenty of support if you run into any issues. Mine went together without any problems, and is currently "making chips." With a product like this, it's only a matter of time before everyone has a CNC mill or router in the workshop. —Brian Graham

Warning Label Generator

Free warninglabelgenerator.com

Do you ever find yourself in desperate need of a properly bureaucratic security notice regarding intergalactic space travel, or the hazard of harboring what look like snowflakes inside your body? Then this site is for you. The array of standard and original symbols and templates just has me itching to hone my skill at making up non-sequitur warning sign zingers, but the intrigue doesn't end there. Joke potential aside, it's a free and lightning-fast way to provide an air of caution to any danger zone you can dream up. —Meara O'Reilly

Bill Byrne is a multimedia artist, educator, writer, and member of the Painful Leg Injuries. billbyrne.net

Brian Graham is a mechanical engineer and robotics coach, and likes his tools a bit too much.

John Edgar Park works at Walt Disney Animation Studios and hosts the Maker Workshop in the upcoming TV series *Make: television*.

Ken Olsen lives and makes in Adair Village, Ore.

Kenny Cheung is an architect and graduate student at the Center for Bits and Atoms at MIT's Media Lab.

Meara O'Reilly is an intern at CRAFT.

Mitchell Heinrich is a design scientist passionate about sustainable technologies and robots that serve cocktails.

Ross Griffith is an engineer and retired professor of textile technology.

Ryan Beacom studies interactive media development at Algonquin College.

Have you used something worth keeping in your toolbox? Let us know at toolbox@makezine.com.

Tick, Tick, Tick ...

The Scenario: You've worked late into the night as a computer engineer in the high-rise headquarters of an international bank, and you're finally heading with your briefcase to your car in the subterranean parking garage. Your car is the only one left on this dimly lit level, parked along a cement wall right near the elevator. But, as you pull out your keys and are about to hit the unlock button, you hear a loud *beep* behind you.

Startled, you turn to see an object against the wall just a few feet away with a pulsing red light on it — and in the poor light, you can immediately make out an illuminated timer *which is now ticking off the seconds from a 3-minute window*!

There is a jumble of multicolored wires, and an array of three motion detectors set to cover a 180° field off the wall, all of which are wired into a small black box sitting on a large brick-shaped object that's slightly smaller than a shoebox. Also atop the brick and on its ends, you see three horizontal glass tubes that appear to contain mercury with wires at both ends, as well as a metallic-looking cylinder with several long wires jammed into the side of the brick-like mass. There's little doubt left in your mind now that *this is a bomb!* — and your arrival here must've set off the timer.

The Challenge: Though you know how mercury switches work, you're uncertain of the purpose of the motion detectors, or of the black box — could it contain a hidden transponder? If you try to move out of range or call for help with your cellphone, might your attempt to flee or the cellphone signal set the device off? Hell, even pushing the unlock button on your key ring now could send the wrong kind of signal, no? But panic is not an option, as it seems you have less than three minutes to decide your best course of action. *So what are you going to do?!*

What You Have: Your briefcase and pockets contain what a computer engineer might normally have, within reason — if that includes a Swiss Army knife or Leatherman tool, so be it. Beyond that, your brain is the best tool you've got. So think fast, and ... good luck.

Send a detailed description of your MakeShift solution with sketches and/or photos to makeshift@makezine.com by March 6, 2009. If duplicate solutions are submitted, the winner will be determined by the quality of the explanation and presentation. The most plausible and most creative solutions will each win a MAKE T-shirt and a *MAKE Pocket Ref*. Think positive and include your shirt size and contact information with your solution. Good luck! For readers' solutions to previous MakeShift challenges, visit makezine.com/makeshift.

And the next MakeShift challenge could be yours! That's right, we're throwing open the doors and offering you the chance to create your own MakeShift to challenge the world. Just submit an original scenario in the familiar format — the challenge, what you have, etc. — with some ideas of how you think it should be solved. The winning scenario will not only be published right here but will also earn you a $50 gift certificate for the Maker Shed. The deadline is March 6, 2009, so get out there and start looking for trouble!

Lee David Zlotoff is a writer/producer/director among whose numerous credits is creator of *MacGyver*. He is also president of Custom Image Concepts (customimageconcepts.com).

Photograph by Jen Siska

Workshop

Making Connections

Neatly lined up along the walls of Len Cullum's 1,500-square-foot north Seattle workshop are handmade Japanese chisels, saws, and planes.

In a building rumored to have once been a shark oil processing plant, Cullum, 42, creates Japanese-style shoji doors and windows, garden structures, and furniture.

Cullum constructs these pieces using traditional joinery, a specialty where it's crucial to be precise and to understand the temperamental qualities of wood because there aren't any metal fasteners to hold together poorly measured or cut pieces. Port-Orford-cedar is his favorite wood to work with. "It planes to an amazing sheen and it smells great," he says.

His philosophy dovetails with his work. "I've long felt that life is largely about making connections. When connecting two ideas or people (or pieces of wood), not only is the fit important, but also the type of connection," Cullum muses. "Some things benefit more from a flexible, freer connection, some from one tighter and more rigid. But none survive one that is sloppy or poorly fit."

Although he uses power tools for the larger cuts, he prefers handmade hand tools for the finer details. "Things that are made by hand have a kind of vibration to them," he says. "The little inconsistencies, even ones you can't consciously see, give it a life."

—*Laura Cochrane*

1. Antique Japanese rip saw used to cut planks from a log. 2. Hand saws. 3. Push chisels used for lighter work. 4. Bench chisels, crane-neck chisels, and shoji-specific chisels. 5. Hollow chisel mortiser to drill square holes. 6. Antique Chinese plane that "takes shavings as thick as cardboard." 7. Japanese hand planes, which are pulled instead of pushed. 8. Japanese hammers used to drive chisels, adjust planes, and tap joints. 9. Cabinetmaker's bench. 10. Parks band saw.

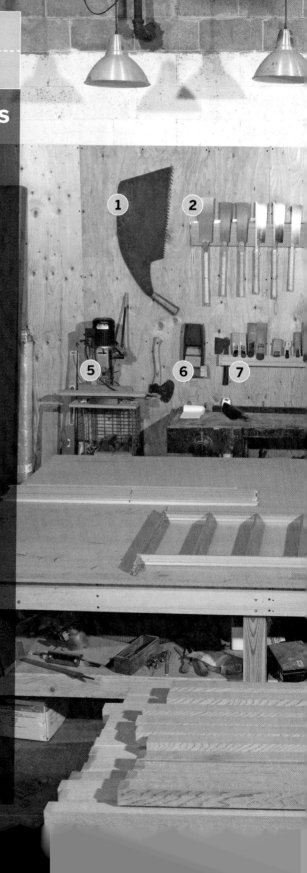

Photograph by John Keatley
» More info: shokunin-do.com
● More images: makezine.com/16/workshop

MAKER'S CALENDAR
Compiled by William Gurstelle

Our favorite events from around the world.

Make: television
Early January, nationwide, U.S.
Catch the first season of *Make:* television. The new series, based on our magazine, comes to public television in early January. Hosted by John Edgar Park. Check local listings for airtimes in your area.
makezine.tv

Photography by Markus Haselbach (bottom) and Matt Blum (top)

» DECEMBER

» Roboexotica Festival for Cocktail Robotics
Dec. 4, Vienna, Austria
The quintessential cocktail robotics event expands its scope as the festival celebrates its tenth year with an exhibition, workshops, musical performances, and an awards gala.
roboexotica.org

DIY Festival of Art and Technology
Dec. 5–7, Zurich, Switzerland
This Euro DIY festival is a series of talks, concerts, and workshops intended for those who prefer to make things their own way. Exhibitions include robots, interactive installations, games, noise toys, and more. diyfestival.ch

» JANUARY

» Fermilab Family Open House
Jan. 17, Batavia, Ill.
The public is invited onto the grounds of the gigantic Fermi National Accelerator Laboratory to see the Cockroft-Walton accelerator, the linear accelerator gallery, the main control room, and more.
ed.fnal.gov/ffse/openhouse

» Techfest
Jan. 24–26, Bombay, India
More than 45,000 people gather at Techfest, one of Asia's largest technology festivals. Activities include several competitions, lectures, workshops, and exhibitions. techfest.org

» FEBRUARY

» Discover Engineering Family Day
Feb. 21, Washington, D.C.
Thousands of kids will visit Washington's National Building Museum to celebrate Engineering Week 2009. The festival features dozens of hands-on engineering-related activities — and slime!
eweekdcfamilyday.org

» Magnet Lab Open House
Feb. 21, Tallahassee, Fla.
The National High Magnetic Field Laboratory opens the doors of its world-class research laboratory at Florida State University for public tours. Demonstrations, giveaways, and the chance to meet and chat with Mag Lab scientists are highlights.
makezine.com/go/magnetlab

» Winter Star Party
Feb. 21–28, Spanish Harbor Key, Fla.
The stars shine bright at this stellar party, where more than 600 amateur astronomers gather to talk about telescopes and gaze at the Orion Nebula and other winter sky attractions without fur-lined parkas.
scas.org

IMPORTANT: All times, dates, locations, and events are subject to change. Verify all information before making plans to attend.

Know an event that should be included? Send it to events@makezine.com. *Sorry, it's not possible to list all submitted events in the magazine, but they will be listed online.*

If you attend one of these events, please tell us about it at forums.makezine.com.

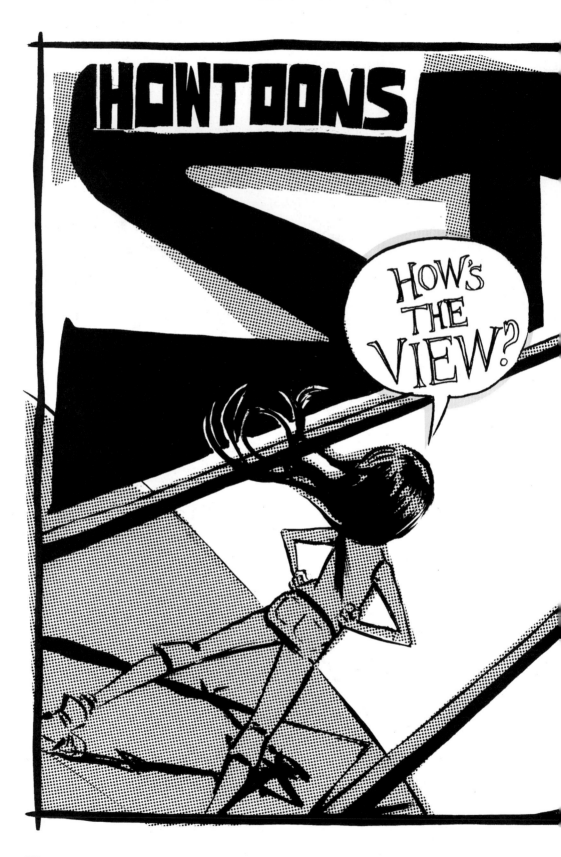

MAKE's favorite puzzles. (When you're ready to check your answers, visit makezine.com/16/aha.)

World Series Survival

The CIA asks you to infiltrate the North Korean leadership. After attempting to enter the country, you're captured by Kim Jong-il. But it turns out he's a fan of American baseball, and as the World Series is about to start, he chooses to give you a sporting chance to be freed.

The Yankees are playing the Red Sox and the 2 teams are perfectly, absolutely matched. Each game is a coin flip — the chances of a win for both are exactly 50%. The outcome of any previous games does not give you any information about who'll win the next game.

The Dear Leader wants to bet a million bucks that the Yankees will win the entire series. He will front you the cash (i.e., he'll give you $1 million up front, and you give him $2 million back if the Yankees win, and give him nothing if the Yankees lose).

If you decline the bet, he will kill you. If the Yankees win and you don't have all $2 million to pay him, he will kill you. If the Yankees lose, and you have money left over, he will accuse you of stealing and kill you. You have no money, except for the money he's given you, and maybe a couple hundreds in your pocket (for rounding errors).

There are bookies you can place bets with in Pyongyang, but unfortunately they will only accept bets for one game at a time, not for the whole series. Since each game is a coin flip and the outcome of previous games doesn't change that fact, you can place whatever size bet you want on each game, at fair odds ($1 wagered wins $2 back), right before they occur (after any previous games have been played). Obviously you must have the money in your pocket to place the bet.

Remember, the World Series is a 7-game series where the first team to win 4 individual games is declared the winner.

How should you place your bets to guarantee you'll have 2 million bucks for Kim Jong-il if the Yankees win the entire series, and $0 if they lose?

Michael Pryor is the co-founder and president of Fog Creek Software. He runs a technical interview site at techinterview.org.

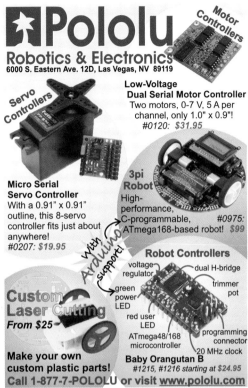

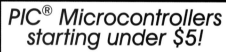

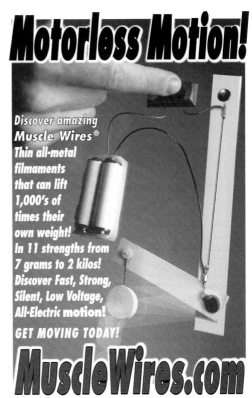

HOMEBREW

Wheelchair Safety System
By Bryant Underwood

■ **My daughter Katy uses an electric wheelchair** and last year she went off to college. Her mom and I were concerned about Katy's safety in navigating the campus — she might get her chair stuck or have some other type of trouble and not be able to get to her cellphone.

So I used a Parallax microcontroller to control a GSM cellphone as a "telematics" system for her wheelchair. Inside the gray box mounted on the back of her chair, I use the phone in speakerphone mode with an external microphone and speaker.

The way I did this was by leveraging the old modem "AT" command set that almost any GSM phone has for legacy control. Using a prepaid phone, I broke it down and connected it to the Parallax Basic Stamp 2 microcontroller and a receiver with two wireless remotes. I used an old Sony Ericsson T226 phone, but I've since learned that the Motorola C168i or T720 would have been much easier.

For operation, Katy keeps one remote in her purse while the other one is attached to her chair. If there's a problem, she can push the buttons on either remote and the device will call a local health care provider, then my mobile phone, and then our home, or with another sequence of button presses it will call campus police. When anyone is reached, the unit acts like a standard speakerphone.

The joy for me is that it's always on and I never have to worry. Plus, at boot-up the BS2 sends a command to the phone to put it in auto answer mode. If we're ever concerned, as a last resort we can call the phone and it will answer in speaker-phone mode without ringing (also good for Dad's concern of boys in the dorm room!). I thought about adding GPS, but my daughter vetoed that idea.

Since Katy's chair is always with her, we know she'll be safe. The alternatives were just not acceptable; even pay services were little more than remote-enabled speakerphones with a required wired connection. With the ease of the BS2, the whole project was assembled over a weekend, including time for coding and debugging. So in a very short time at home, I was able to develop and deploy a one-of-a-kind wireless product that exactly fits my needs!

Bryant Underwood serves as the director of supply chain for a defense contractor and resides in the Bridgeport, Texas, area.

Photograph by Debra Underwood